WHY AREN'T
BLACK HOLES BLACK?

THE
UNANSWERED
QUESTIONS AT
THE FRONTIERS
OF SCIENCE

ROBERT M. HAZEN
with MAXINE SINGER

Foreword by
STEPHEN JAY GOULD

Anchor Books
Doubleday

NEW YORK LONDON TORONTO SYDNEY AUCKLAND

WHY AREN'T

BLACK

HOLES

BLACK?

AN ANCHOR BOOK
PUBLISHED BY DOUBLEDAY
a division of Bantam Doubleday Dell Publishing Group, Inc.
1540 Broadway, New York, New York 10036

ANCHOR BOOKS, DOUBLEDAY, and the portrayal of an anchor are trademarks of
Doubleday, a division of Bantam Doubleday Dell Publishing Group, Inc.

Library of Congress Cataloging-in-Publication Data
Hazen, Robert M., 1948–
    Why aren't black holes black?: the unanswered questions at the
  frontiers of science / Robert M. Hazen and Maxine Singer.
      p.   cm.
    1. Science—Miscellanea.   I. Singer, Maxine.   II. Title.
Q173.H42   1997
500—dc21   96-29856
                                                                     CIP

ISBN 0-385-48014-8

Book design by Brian Mulligan

10 9 8 7 6 5 4 3 2

To our friends and colleagues of the Carnegie Institution of Washington, in admiration of their ongoing pursuit of the unanswered questions

Contents

The Frontier Remains Endless

Stephen Jay Gould

When I was a kid on the streets of New York, I had two consuming passions: stickball (combined with active rooting for the New York Yankees) and science (combined with personal dreams of becoming a paleontologist "when I grow up"). The Yanks did very well during those years, and I did become a paleontologist. Sometimes, from my current perspective in mid-life, I look back at those dreams and pinch myself for reassurance about the current reality. I thought, when I was ten, that science would provide a life of continuous excitement and fulfillment, but what did I know? No one in my family had taken such a route; few had ever gone to college. Sometimes our dreams pop like a balloon against the rough wall of reality. (My uncle Mordie was a wonderful violist, and played first chair nearly forever in the Rochester Symphony. When I asked him at retirement how he had enjoyed life in such a profession, he replied: "It was fine, except that I hate music.") But science turned out to be all that I had hoped, and more. I remain as thrilled with my decision long fulfilled as I ever was with my childhood's distant dream and idealization.

I could list many reasons for this continuing sense of privilege and reward, and I could also document some frustrations (the hours are truly terrible—at least thirty-six each day when you're on to something big). But I think that the greatest satisfaction was captured particularly well in the title to Vannevar Bush's famous report on the state of American science after the Second World War: *Science: The Endless Frontier.* Science is the best technique we have ever developed for gaining factual knowledge about the empirical world. But any good scientist has always learned, even after a triumphant lifetime filled with personal discovery, that we have mastered only the most pitiful fraction of what might be known—and that, to make the point even more emphatic, any additional knowledge reveals an even larger domain of unknown things that we hadn't even been able to conceptualize before. This sense of an endless frontier has been embodied in many famous aphorisms by scientists and about science. I particularly like Pascal's comment that the growth of knowledge is like an expanding sphere in space: The greater our understanding (symbolized by the sphere's volume), the greater our contact with the unknown (the sphere's surface).

Two critiques of this view—exactly opposite in their claims but entirely similar in their denigration of the future of science—have attained some popularity in recent times. The first, as stated in extreme form by some social critics of science, denies that science (or any other human enterprise) can attain truly objective empirical knowledge about the external world. This extreme form of relativism holds that although the objective world exists, human observations must be so enmeshed in biases—of social and political contexts, of personal psychology, of culturally imposed gender roles, etc.—that all knowledge becomes more of a social construction than a verifiable empirical reality. In this extreme view, the history of changing scientific ideas doesn't record any closer approach to empirical truth, but only the shifting tides of social preferences.

This line of argument represents a good point taken so foolishly far that it ends up as perniciously wrongheaded. All science (and all scholarship of any sort) must be embedded in biases of the sort mentioned above. But this observation only acknowledges the human side of the scientific enterprise, and should be welcomed by all practicing scientists, if only as a way of lessening the fear and incomprehension that so often defines the general public's attitude toward science. If everyone understood that scientists do their work surrounded by all the foibles of any creative human enterprise, the profession would seem less like an inaccessible priesthood, and more within the grasp of any intelligent reader.

Yes, science must proceed in a social context that biases the results. But science also moves—however fitfully, full of temporary stallings, and even occasional backward plunges—toward better understanding of the external world, and I do not know what we can call such a process except a general increase of true and objective knowledge. (Moreover, if scientists accept the necessary social bias in their work, they can approach their research with more vigilance and scrutiny, therefore becoming better able to separate the empirical signals from the cultural overlay.) I hate to make such an old-fashioned, practical (one might almost say philistine) argument, but I know no better evidence for the progress of science by increasing understanding of the empirical world than the obvious facts of our technical success. How could we have reached the moon or learned how to read the human genome if the history of our field only recorded a set of shifting social constructions?

The second, and opposite, argument holds that science has become the victim of its own spectacular success, and has now reached an effective termination where nothing interesting remains to be done—thus relegating the enormous cadre of contemporary scientists to the comparatively dull task of filling in the few blanks of an essentially completed theoretical structure. Proclamations of the end of any enterprise always make good journalistic copy, and a virtual industry of such punditry exists, with announcements of the end of almost anything you can imagine, from ideology, to history, to pole vaulting (ah, that fiberglass pole) or baseball (ah, that aluminum bat). As I write this introduction in the late summer of 1996, John Horgan's astonishingly superficial book on the end of science is all the rage among science's critics and commentators, and a source of infinite boredom among practicing scientists.

I could now unleash a passionate disquisition on the current state of our profound ignorance about many of the deepest questions, and the consequent impossibility of any imminent termination (unless all these questions are unanswerable in principle, as they are not). But I desist because this book, *Why Aren't Black Holes Black?* with its careful exposition of what we do not know (in the joyful context of the great amount we have already learned), provides the finest antidote for such nonsense. I note only the absurdity of talk about the end of science when the leading theory of cosmology can present evidence for a mere ten percent of the matter required to make the scheme work; and where we have only the vaguest clues for a proper theory of memory and the basic structure of consciousness.

Our blessed (and correctable) ignorance is entirely fractal, and operative at all scales. I shall cite just two personal examples at opposite ends of generality. From the smallest arena of technical detail, my own research concerns the evolution of Caribbean land snails during the last 200,000 years. When I began these studies in the 1960s, we were stymied by a lack of any technique for determining radiometric ages in this interval. Radiocarbon has too short a half-life (between five and six thousand years), and can date material back to only forty or fifty thousand years at most. The other techniques then available used isotopes with too long a half-life (often millions or billions of years), and therefore couldn't date rocks between 100,000 and 200,000 years old for the opposite reason of insufficient resolution. We were absolutely stuck and thus enormously frustrated. I simply couldn't pursue many fruitful avenues of study that seemed so ripe for important conclusions. Since then, a variety of new techniques have filled in the gap, and I can accurately date sediments of all ages in my study area.

At the largest scale of general interest, I write this foreword just two weeks after the publication of an article announcing the discovery of probable life at bacterial grade in 3.6-billion-year-old carbonate infillings of a Martian meteorite. Although the data may not pan out, the case is eminently plausible because Mars maintained sufficiently earth-like conditions (with running water) at the time, and because the Earth did evolve life at such a grade during this early period of our planet's history. The Martian surface almost surely does not maintain appropriate conditions for life today. But subsurface rocks on Mars (where enough water may percolate) could continue to support life, just as bacteria thrive in pore spaces of subsurface rocks on Earth today. We can—and presumably will within the next two decades—send a lander to Mars and return with robotically collected subsurface samples.

If we find life in such Martian rocks, we will have potential evidence for a kind of "holy grail" issue that we cannot even approach today. (And so much for the supposed end of science!) All living things on Earth share an astonishing set of precise biochemical similarities (building of genetic codes by nucleic acids, storage of energy in ATP, etc.).

We yearn to know whether these similarities represent the only way to make life (and must therefore serve as the biochemistry of living things wherever they evolve), or whether earthly similarities represent the one solution that happened to arise among numerous workable alternatives (therefore implying that life independently evolved elsewhere might use a very different biochemistry). We cannot now approach this question because all life on Earth represents only a single "experiment"—for all living things on Earth have probably descended from a unique origin. We must have another "experiment" to begin to answer this question of questions—and Mars represents a real possibility, potentially available in our lifetimes. I would dearly love to know before cashing in my earthly chips.

The sources of our ignorance are many, but all correctable. Some just represent an absent technology to resolve issues that we know how to formulate (my two examples at opposite scale—Martian life, and dating rocks in the 100,000-to-200,000-year range—belong to this category). Other sources, in many ways more interesting, represent failure of our own intellectual equipment—for we often do not even know how to think about issues now shrouded in mystery. Some of these limits may reflect simple stupidity on our part, but others probably record limits imposed by the structure of our brain as an evolved organ for contexts so different from present needs and uses. We are, for example, very bad at reasoning about probabilities. We also have an almost overwhelming tendency, often not very useful these days, for thinking about issues as dichotomous divisions (night and day, male and female, nature and culture). I do not think that these mental preferences are simple impositions of culture. If we understood how evolution has structured our preferred ways of thinking, we might be able to transcend these limitations and break barriers by internal reordering rather than technological invention.

When I finished playing stickball and reading my books about dinosaurs, I would often lie awake in bed at night and, in my aimless eight-year-old way, try to resolve the unimaginable issues of eternity and infinity. How can the universe have an end? Well, maybe if you keep going, you run into a brick wall. But what's on the other side of the brick wall? And so on. I don't think that we shall soon encounter any brick wall in our paths of scientific discovery. It ain't over until it's over. And as for the fat lady (an operatic image, I assume, and therefore particularly appropriate because Bob Hazen is one of the nation's best baroque trumpet players as well as a fine scientist)—well, she has hardly begun to warm up. Someday we might give her an encore: "Glory, Hallelujah. Truth is marching on."

Museum of Comparative Zoology
Harvard University
Cambridge, Massachusetts

PREFACE

To know that we know what we know, and that we do not know
what we do not know, that is true knowledge.
                                   —Henry David Thoreau, *Walden*

In 1831 a young graduate from Cambridge University set out on an epic
five-year voyage of adventure and discovery—a journey of body and
mind that was to change forever our perceptions of ourselves. Though
far from a fully trained naturalist, Charles Darwin left his native England to serve as geologist and biologist for the ambitious South American voyage of the H.M.S. *Beagle*. Through dangers and hardships, he
recorded details of the plants, animals, and fossils of distant jungles and
remote islands. All the while, he became increasingly captivated by one
of the great unanswered questions of his day: How did the extraordinary
diversity of life on Earth arise? Darwin's intense curiosity, channeled by
observation, reasoning, and remarkable insight, ultimately led to his
sweeping theory of evolution by natural selection.

All scientific research, like Darwin's journey of discovery, is an outgrowth of our insatiable, uniquely human curiosity. A myth has arisen
in our time that scientists have all the answers. What may, perhaps, be
true is that scientists are more acutely and joyously aware of how much
they *don't* know, and they are relentlessly driven to seek explanations
for what they don't understand. The quest for answers may proceed
with meticulous care or reckless flamboyance, in precise logical steps or
wildly intuitive leaps. Advances are made with a computer terminal, on
the lab bench, at billion-dollar machines, or in the shower. Some of its
practitioners collect data sitting quietly at a microscope in a darkened
room, while others risk their lives at ocean depths or on the slopes of
active volcanoes.

Knowing

It is in vogue in some academic circles to debate the nature of reality and what it means "to know." In this spirit, historians and philosophers of science of recent decades have focused attention on the nature of knowledge itself. They ask if scientific truth is absolute, whether a scientific theory can be proven, and how new theories replace or subsume older theories. They argue about whether scientific knowledge is progressive, and debate the extent to which its pursuit is rational.

Most working scientists find the questions posed by philosophers of science mildly intriguing, occasionally annoying, and largely irrelevant to their day-to-day efforts. As researchers who have worked in the trenches for decades, we believe that the only practical way to understand how the universe works is to ask questions that can be answered by making observations, performing experiments, and devising models that are independently reproducible and verifiable. We believe, perhaps with naive optimism, that many attributes of the physical universe can be known with certainty, and that the scientific process of observation and experiment is the best way to discover these traits. Given that premise, which is shared by the vast majority of people who do research for a living, we recognize three broad categories of factual knowledge.

The first category encompasses facts that we know to be true based on reproducible observations. We know the Earth is round (sort of), that light travels 186,000 miles per second in a vacuum, and that living things on Earth are formed from carbon-based molecules. Our cumulative scientific knowledge is vast—much greater than any individual could grasp.

A second category of knowledge includes everything that we know we *don't* know. To be sure, much of this unknown information is trivial and uninteresting. It's doubtful that anyone has bothered to determine the electrical conductivity of peanut butter or the effects of AM radio waves on amoebas, although the measurements could be made easily enough if we wanted to. But much of what we know we don't know is profoundly exciting and important to obtain. Finding a cure for AIDS, learning the causes of aging, or discovering an unusual new life-form will change our world and influence our lives. Most discoveries, furthermore, open new doors to scientific inquiry. The discovery of the structure of DNA four decades ago, for example, was one of the most widely recognized and passionately pursued problems in science. While few scientists could have anticipated the revolution caused by molecular genetics, many knew that any deep understanding of life had to wait for the breakthrough discovery of what we now know as the double helix.

Finally, there exists a vast store of knowledge that we don't know we don't know. One of the most exciting aspects of science is the discovery of completely unexpected new objects and phenomena that no one could have imagined. X rays, viruses, lasers, polymers, superconductors, dark matter, distant galaxies—the list of amazing discoveries in the last century alone is astonishing. How many more extraordinary discoveries await scientists and science watchers? We have no way to predict the nature or impact of such discoveries.

The Unanswered Questions

The physical universe is like an amazingly complex cavern. It exists in its entirety, with numberless rooms and corridors, vaulted chambers and barely passable cracks, all interconnected in incomprehensible dimensionality. Many rooms, some of awesome grandeur and others of intimate beauty, are well traveled. Every stalactite and stalagmite has been meticulously measured, every ripple on the aged walls described. Other passages and openings have been roughly mapped and merely await the careful observer to complete their documentation. Some scientists choose to spend their lives in these safe, secure areas, where there is little chance of getting lost, and even less chance of getting noticed.

But the glorious cavern that is the physical world holds great mysteries as well. Some researchers attempt to understand how the cave came to be. They probe the origins of the chambers and their graceful formations, and the processes by which these features change. They seek to explain the known parts of the cave, and predict what will be found in regions as yet unknown.

Others attempt to unlock secrets of the vast unexplored regions. Some enticing corners of the cavern are dimly visible across gaping chasms; these recesses await new technologies to be conquered. Some magnificent chambers are completely hidden and thus remain undiscovered. Exploration of the maze is undertaken with considerable effort and frustration, for many uncharted passageways end abruptly or twist and turn back to the starting point. Life is too short to spend it in aimless wandering. But once in a while, if well armed with training, vigilance, and luck, an explorer might happen upon some new vista of such sweep and magnificence that it will forever change our view of the world.

Why Aren't Black Holes Black? examines the limits to which we have probed the natural world and celebrates daring efforts to go even further—a saga of the most exciting and mysterious galleries that lie just beyond our reach. In selecting the unanswered questions of science for inclusion in this book, we have tried to document the broad scope of what scientists are actually studying today, as well as what they hope to learn in the years to come. In so doing, we have neglected many profound questions that we can ask but that science is not yet in a position to answer.

Ask any scientist to name the most important unanswered questions in science, and a few topics will probably appear on most lists: the origin of life, the nature of matter and forces, and the ultimate fate of the universe, for example. Other key questions included in this book are dictated by the vast research efforts now being devoted to them. Countless thousands of scientists around the world tackle questions related to new materials, energy, aging, and the environment. Inevitably, the list reflects our own interests and prejudices. Chapters on the Earth's deep interior, dark matter, developmental biology, and genetics relate directly to our own research and that of our colleagues at the Carnegie Institution of Washington—work we do because the underlying questions are fascinating and, we believe, important. Indeed, any scientist who does not include his or her research field in a list of the top unanswered questions should consider switching fields.

Several high-profile topics are not included in our list of key unanswered questions. The origin of the universe, the paradoxical quantum nature of matter, and the complex structure of the cell would have figured prominently several decades ago, but great strides have been made in understanding these subjects. New questions about the universe, matter, and life have taken their place. Some authors might have included the emerging (and much hyped) field of complexity and the closely related studies of chaos and information theory as a separate inquiry. While these subjects appear in several chapters, the ambitious search for a "unified theory of complex systems" seems too vague and ill focused in its present state to represent one of the great unanswered questions of science.

Clearly, a survey of the unanswered questions in science today can focus only on those aspects of the universe that we *know* we don't know. The most exciting part of the scientific drama, however, is the continual discovery of phenomena that we *didn't know* we didn't know. The list of key unanswered questions will change dramatically in the years ahead as new, now unimaginable questions come to light. Despite facile pronouncements of the "end of science," the chain of discovery—and human curiosity that drives our search—shows not the slightest sign of ending.

We have intended to write a general overview of today's most compelling questions in the physical and life sciences rather than offer exhaustive coverage of any one scientific topic. Each chapter summarizes a rich history of discovery, and explores the ongoing work of countless researchers. Readers interested in learning more about these topics are encouraged to examine the additional sources at the end of the book. And keep an eye on the newspapers! Science discoveries are part of the current world adventure in which everyone can share.

WHY AREN'T
BLACK HOLES BLACK?

Prologue:
The
Nature of
Questions

It is better to know some of the questions than all of the answers.

—James Thurber

Science, first and foremost, is a search for answers. But every answer must begin with a well-conceived question. Too often we focus on the most firmly established of scientific answers—the most significant discoveries scientists have made about the physical world. In the process, the dynamic, creative process of scientific inquiry is sometimes shortchanged.

Not all questions are equal. They differ in significance and in the scope and complexity of their answers. Here are just a few of the characteristics of inquiry that shape and inform the scientific process.

• Many Important Questions Are Beyond the Realm of Science.

Science addresses only those questions that can be answered by reproducible observations, controlled experiments, and theory guided by mathematical logic. This distinction between scientific and nonscientific inquiry, though sometimes blurred, is neither frivolous nor arbitrary. Science can reveal if a painting is old, but it cannot determine if the painting is beautiful. It may be used to deduce the origins of the physical universe, but it cannot rationalize why we are here to ponder its existence.

Many of the most important questions we face—What is the meaning of life? Whom should I marry? Is there a God?—lie outside the domain of science. Such a realization led economist and philosopher Kenneth Boulding to remark, only partly in jest: "Science is the art of substituting unimportant questions that can be answered for important questions that cannot."

• Scientific Questions Are Richly Varied in Scope and Content.

Scientific inquiry is as diverse as the natural world itself, but most scientific questions fall into four broad categories:

1. *Existence Questions*. These are questions that ask "What's out there?" Scientific explorers of past centuries reveled in voyaging to exotic lands in the pursuit of animal, plant, and mineral specimens. Chemists isolated element after element, physicians dissected diseased corpses, astronomers catalogued countless stars, and physicists scrutinized unusual phenomena associated with electricity and magnetism. Ongoing efforts to observe galaxies, sequence genomes, isolate viruses, unearth dinosaurs, and eavesdrop on extraterrestrial life are all attempts to answer the question "What's out there?"

Naming and categorizing the natural world continue to be essential facets of scientific inquiry. Their answers may reveal underlying patterns in the natural world and provide the basis for further exploration.

2. *Origin Questions*. These explore how natural objects and phenomena came to be. Questions about the origin of the universe, the Earth, and life itself are among the most fascinating mysteries of science. They must be answered with a plausible sequence of events, deduced from contemporary evidence and constrained by natural laws.

3. *Process Questions*. Such questions are often closely linked to questions about origins. They explore how nature works: how stars evolve, how rocks erode, how cancer develops, how atoms interact, and how fungi reproduce. Descriptions of the dynamic interplay and evolution of natural systems help us to understand the past and present, and to predict the future of our physical surroundings.

4. Applied Questions. These questions look for ways to manipulate the physical world to our advantage, whether curing disease, devising new materials, or modifying the environment. Such questions are often quite specific and rooted in technology: the search for safer cars, cleaner energy, cheaper food, and longer lives. Such pursuits are closely tied to other types of scientific questions, for we can't manipulate nature without first understanding how it works. Curing infectious diseases, for example, requires a knowledge of infectious organisms and how and why they behave the way they do; construction of a skyscraper relies on a detailed knowledge of the properties of materials.

• Some Questions Are Not Now Scientific, but May Be Someday.

Before Edwin Hubble's discovery of distant galaxies, the question of how the universe began lay outside observational science. Without relevant data, it was a matter of philosophical speculation. Similarly, "what is time?" is today more a question of philosophy than of science, though that situation may change as we learn more about matter and energy.

• Most Questions Cannot Be Answered Completely.

We are prevented from obtaining complete answers to many scientific questions. Among these limitations are:

1. Experimental Error. All measurements, no matter how accurate, contain some error. Science is a never-ending pursuit in part because new techniques for making observations and measurements—bigger telescopes, faster electronics, more sensitive analytical methods—regularly improve available accuracy and inspire new experiments on old questions.

2. *The Uncertainty Principle*. At the subatomic scale, every measurement alters the object being measured. One of the most unsettling discoveries of quantum mechanics is that at nature's smallest scale, matter and energy come in discrete packets called quanta. Any observation of a quantized object (for example, a measurement of its velocity) causes it to change (i.e., its position). Thus, every observation of a natural phenomenon has a built-in uncertainty at the atomic scale.

3. *Chaos*. Many natural systems are chaotic, and thus are inherently unpredictable. Isaac Newton's laws of motion seem to imply an orderly universe of cause and effect in which any given set of initial conditions leads to a predictable outcome. Recently discovered "chaotic systems," including dripping faucets, human heartbeats, and the weather, obey Newton's laws, but the most minuscule variation in their initial conditions can radically alter their behavior. Since every measurement of initial conditions contains some error, the future behavior of chaotic systems cannot be predicted accurately.

4. *The Speed of Light*. Limitations are placed on us by space and time. The cosmic speed limit of 186,000 miles per second—the speed of light—prevents us from knowing the state of distant galaxies at this moment, while the forward direction of time prevents us from visiting the past.

In spite of these inherent limitations on inquiry into the natural world, the methods of science provide the most effective and powerful tool we have to understand and modify our physical world.

• Some Questions Are More Fundamental Than Others.

Some research objectives are more profound than others. Research programs that target key steps in the origin of life are more likely to provide insight than exhaustive surveys of exotic plant and animal species. Of course, while a seemingly routine observation or experiment is usually just that, one never knows when a startling new result might be found.

Scientists gain deep insight from all four broad types of scientific questions. Answers to the existence of extraterrestrial intelligence, the origin of life, the process of evolution, and the preservation of the environment would all constitute major scientific advances. We must also remember that the complexity or length of an answer is not a measure of a question's significance. An unambiguous yes or no answer to the question of whether intelligent life exists on other worlds would be as profound as any human discovery in history, while an exhaustive multivolume catalogue of New Jersey's rocks and minerals would likely provide few fundamental new insights.

Profound questions are not always obvious questions. While the birth of the universe, the origin of life, and the inevitability of aging and death have invited speculation for thousands of years, other compelling questions, such as the nature of energy, the control of genes, and the mystery of dark matter, are far more subtle, emerging from the nagging persistence of odd observations and anomalous bits of data. Gradually, over the span of decades or even centuries, we become aware of a fundamental lack in our understanding of the physical world, and a deep mystery—a new question—comes fully to light.

While we have no absolute method to quantify the relative significance of different questions, a few thinkers have been emboldened to offer their personal lists of the most fundamental questions. Physicist Paul Davies in *God and the New Physics*, for example, examines what he calls "the Big Four questions of existence:"

> Why are the laws of nature what they are?
> Why does the universe consist of the things it does?
> How did those things arise?
> How did the universe achieve its organization?

In a similar vein, astronomer Alan Dressler has defined "three basic questions of science:"

Where are we?
What are we?
How did we get here?

As sweeping and broad as these questions are, they are at least one step removed from the kind of inquiry that scientists can pursue in any systematic way. These are great unanswered questions, but they are not phrased in a way that allows us to perform an experiment or devise a theory. Davies's and Dressler's questions lack the sharp focus necessary to organize effective research (or obtain a government grant).

Advancement in science often arises from one's ability to shed light on a big question by tedious and exacting pursuit of a small one. Charles Darwin developed his sweeping theory of evolution by meticulous examination of living and fossil animals, Galileo Galilei deduced a comprehensive theory of motion by watching balls roll down an incline, and Edwin Hubble discovered the expansion of the universe after measuring the light from many distant galaxies. Scientists cannot avoid the rigors of exacting observation, experiment, and theory. Still, inspired intuition often provides a shortcut from esoteric detail to profound insight.

The question of relative importance of scientific endeavors comes most forcefully into play in the day-to-day world in the awarding of prizes and research grants. Academic prizes and the prestige associated with them are usually conferred for work perceived as fundamental and basic. Ernest Rutherford's discovery of the atomic nucleus, Linus Pauling's explanation of the chemical bond, and James Watson and Francis Crick's unraveling of the structure of DNA were all recognized by Nobel prizes. Research grants, on the other hand, are usually awarded by government agencies—the departments of Commerce, Health and Human Services, Defense, or Energy, for example—that have specific pragmatic goals in mind. Many scientists, accordingly, become adept at rationalizing their pursuit of basic research by demonstrating how this research is essential to solving applied problems.

This dilemma is well illustrated by the current efforts to find a cure for AIDS. Basic discoveries related to viruses, DNA, and the immune system, for example, may not directly benefit anyone infected with

HIV, but they will be essential to finding a cure for AIDS and other diseases. Discovery of an AIDS vaccine, by contrast, may not result in any fundamental new understanding of biological systems, but will have an immediate and profound effect on millions of people. Any AIDS-research funding strategy must thus strike a balance between basic and applied efforts.

The relative importance that we attach to questions, however, is highly subjective. Every culture asks different questions, reflecting their varied beliefs and experiences. All cultures wonder at one time or another about the vast scale of the universe and the ancient origin of life. Not all cultures, however, ask how many stars exist or the exact age of fossil species, much less how these objects evolve.

Scientists inevitably bring personal biases to research topics. Research on conservative treatments for breast and ovarian cancer lagged far behind those for testicular cancer for most of this century, until women began to represent a significant percentage of the biomedical research community. Indeed, the first experiments on nonsurgical treatment of breast cancer were undertaken in Italy on wives of doctors. Gender bias may also be perceived in the story of geneticist Barbara McClintock's discovery of the extreme plasticity and mobility of genetic material, which took twenty years to be recognized by the scientific establishment.

Such gender differences may be deeply ingrained and influence the choice of research questions. Gerhard Sonnert and Gerald Holton of Harvard University report in a 1995 study that half of male scientists and almost two-thirds of women scientists believe that men and women do science differently. As the scientific enterprise becomes more accessible to women and minorities from around the world, the scope and character of scientific questions will inevitably become enriched with new perspectives and insights.

• Scientific Questions Are Often Interconnected.

As scientists explore the most sweeping unanswered questions, they often discover links between what seem at first to be unrelated topics. Theories of the subatomic structure of matter have informed questions of the origin and ultimate fate of the universe. Plate tectonic models of the Earth's dynamic interior bear directly on our understanding of life's origin and evolution. Principles developed in the search for new materials may shed light on the origin of the first living cell. And studies of ancient mass extinctions provide models for understanding the importance of the global environment. Big questions tend to blur the traditional boundaries of science departments and unify the study of our world.

The approach to answering some questions, such as the dark-matter puzzle, the search for a theory of everything, and the study of genetic influences on behavior, is based on the assumption that systems can be understood by examining the behavior of fundamental building blocks. At the opposite extreme, ecosystems, the brain, and the universe as a whole can be examined as collective systems that display properties completely unlike those of their smaller components. Some big questions, including those related to the Earth, the brain, and the environment, focus on the often chaotic way complex systems evolve. These recurrent themes in scientific questions provide new ways of approaching old problems and establish links among seemingly unrelated disciplines.

• Answers to Old Questions Often Lead to New Questions.

In 1843, art critic John Ruskin lamented, "To know anything well involves a profound sense of ignorance." This sentiment strikes a responsive chord with most scientists today. The more our knowledge grows, the more we realize how much we don't know.

Black holes provide a good example. When first proposed in the late 1930s, black holes were a startling and strange concept. They were portrayed as the ultimate cosmic vacuum cleaners, with gravitational pulls strong enough to suck up all matter *and light* that came too close. Nothing could escape; nothing but a black void would be seen by the passing interstellar traveler. But further calculations revealed that black holes *could* be seen by the way they bend light coming from more distant stars and galaxies, and a new way to study the universe had been found. Then, in 1973, Stephen Hawking demonstrated that a strange

quantum mechanical property of black holes causes them to spontaneously emit radiation—black holes aren't black after all. Now these
properties of black holes are being exploited by a new generation of
astronomers to answer questions about the nature of matter and the
ultimate fate of the universe.

Many other unanticipated discoveries such as X rays, tectonic plates,
the genetic code, and buckyballs also opened up vast new areas of
research. Even seminal discoveries that appear to wrap up loose ends—
the periodic table that systematizes dozens of chemical elements, the
theory of natural selection that links all life-forms to a common ancestry, and the standard model of high-energy physics that unifies matter
and forces in one comprehensive framework—raise new and deeper
questions about the underlying structures of nature.

We almost certainly have yet to recognize and ask many of the most
profound questions about the universe. As William Harvey observed in
the dedication to his great 1628 treatise on the circulation of blood,
"All that we know is still infinitely less than all that remains unknown."
The true measure of scientific progress is thus not so much a catalogue
of those questions we can answer as it is the list of the questions we
have learned to ask. And for as far ahead as anyone can foresee, there
will be no end to the questioning.

Dark Matter:
Where
Is the
Missing
Universe?

We know our immediate neighborhood rather intimately. With
increasing distance, our knowledge fades, and fades rapidly.
Eventually, we reach the dim boundary—the utmost limits of our
telescopes. There, we measure shadows, and we search among
ghostly errors of measurement for landmarks that are scarcely more
substantial.

—Edwin Hubble (1921)

It wouldn't surprise me to find out that 90 percent of the material
of the universe is missing. I'm sure 99.9 percent of *my* things are
missing.

—Roger L. Welsch (1996)

A huge chunk of the universe is missing. Within the past two decades
astronomers have discovered overwhelming evidence that the universe
is littered with dark matter—seemingly invisible stuff that must be out
there but that we can't find with our most powerful telescopes. Today
no scientific question holds more mystery or significance than the puzzle over dark matter.

Why should we care about matter no one can see? First, because the
most basic task of science is to catalogue and describe the universe
around us. If as much as 99 percent of the universe is unseen and
hidden, as recent evidence suggests, then we have barely begun to document its contents. Second, evidence suggests that the missing mass is
intrinsically different from everyday matter. Our understanding of the
universe and its origin may be woefully biased and incomplete until we
account for dark matter. Finally, the amount of missing mass is closely
tied to the ultimate fate of the universe—whether cosmologists expect
it to expand forever or eventually collapse into itself. The missing mass
problem thus lies at the heart of our most fundamental attempts to
understand the past, present, and future state of the cosmos.

Observing What's Out There

For centuries the central objective of astronomy has been to identify and explain all the objects that make up the universe. For the most part, this effort is laborious and routine. Thousands of astronomers have spent night after bone-chilling night charting the visible universe.

The challenge of documenting the night sky is not unlike the task of entering a room and having to describe all its varied furnishings. First, we observe light coming from every object in the room. The distribution of light defines the location, size, and shape of the furnishings, while distinctive patterns of light absorption indicate whether they are made of wood, metal, plastic, or other materials. Similarly, astronomers measure light from objects in space to deduce their locations and chemical makeup.

Additional information about a room's furnishings—their hardness, for example, or how much they weigh—is gained by measuring how they respond to forces. If we bump into a lightweight folding chair it will easily slide out of the way, but bump into a massive oak desk and it may bruise your shin. Astronomers can't bump into distant stars, but they can observe how stellar motions are perturbed under the influence of gravitational forces exerted by neighboring stars or galaxies.

Time after time, new technologies have allowed astronomers to see and measure fainter and more distant objects. In the early seventeenth century, Galileo's simple telescope first revealed the moons of Jupiter, the rings of Saturn, and countless thousands of stars too faint to be seen by the naked eye. Early this century, Edwin Hubble employed the newly constructed 100-inch telescope of the Mount Wilson Observatory to document thousands of unknown galaxies. And today's astronomers harness the unprecedented power and clarity of the Hubble space telescope to image the most distant visible objects in the universe. These observations have given astronomers the opportunity to answer one of the most basic questions about the universe: how much mass does it contain?

Weighing Galaxies

Almost all the visible matter in the universe is found in galaxies, which exist on a scale almost beyond comprehension. Each galaxy holds tens to hundreds of billions of stars in a region that may exceed a hundred thousand light-years in diameter. (A light-year, the distance light travels in a year, is almost six trillion miles.) Our own galaxy, the Milky Way, contains all the familiar stars and constellations, but billions of other galaxies are visible in powerful telescopes. For astronomers who want to study the nature and distribution of the universe's mass, galaxies are the logical place to start.

Astronomers rely on two complementary methods to estimate a galaxy's mass, in much the same way that two complementary approaches can be used to document a room's furnishings. The quick and easy way is to count the total number of visible stars (an effort greatly simplified by image-processing computers), and then multiply that number times the average mass per star (a number that has taken astronomers many years of observations and theoretical calculations to determine). This calculated value is known as the visible mass of a galaxy.

Alternatively, astronomers determine the dynamical mass of a galaxy by observing how stars move. Powerful gravitational forces hold stars, planets, and the many other pieces of a galaxy together. Gravity keeps these objects from drifting apart into the vastness of intergalactic space, and controls their every motion. Each star or cloud of gas in a galaxy orbits about the center like an immense pinwheel, according to Newton's law of gravity. Newton realized that every orbiting object, such as the moon circling the Earth, experiences a precise balancing of forces. On the one hand, the Earth's gravity constantly pulls on the moon, causing it to fall downward. At the same time, the moon is hurtling outward into space. The moon's orbit at 286,000 miles is a precise balance between its downward fall and its outward motion.

If we could magically increase the Earth's mass, its gravitation pull would increase and the moon would begin to fall faster, adopting a lower, faster orbit. A less massive earth with weaker gravity, on the other hand, would allow the moon to fly outward and adopt a higher, more leisurely orbit. Astronomers use data on orbital speeds and radii to calculate the mass of many celestial objects. They determine the sun's mass, for example, by measuring the length of Earth's year and its distance from the sun.

In much the same way, astronomers determine the dynamical mass of a galaxy by measuring the position and orbital speed of its stars or clouds of gas as they circle the galactic center. The more massive the galaxy, the faster its stars must travel in their galactic orbits. Ultimately, if we have properly accounted for all of a galaxy's variables, the visible mass should exactly match the dynamical mass.

Measuring the Speed of Stars

Calculations of a galaxy's dynamical mass depend on accurate determinations of star speeds. The relative motions of individual stars in a distant galaxy are much too small to detect from Earth's vantage point, even in several human lifetimes of astronomical observations. For all intents and purposes, distant stars are fixed points in the heavens. To measure star velocities in distant galaxies, astronomers must resort to the same technique the police often use to catch speeders—the Doppler effect.

The Doppler effect occurs whenever a source of waves, such as the sound waves from a loud vehicle, moves relative to an observer. As the vehicle approaches, a high-pitched sound is heard because approaching sound waves pile up. As the vehicle passes by, the pitch rapidly drops and a lower pitch is heard because receding sound waves are spread out, lowering the apparent frequency.

Police use the Doppler effect by reflecting a microwave beam of an exact frequency off an approaching automobile and measuring the shifted frequency of the returning beam. The faster the car moves, the greater the telltale frequency shift.

Moving stars also emit a broad spectrum of electromagnetic waves that reveal their velocities. Our eyes detect a narrow portion of this radiant energy as visible light, but stars also emit radio waves, microwaves, ultraviolet radiation, X rays, and gamma rays. Every portion of this rich spectrum of electromagnetic radiation travels at light speed, 186,000 miles per second, in the form of waves. Much of a star's radiation is concentrated in narrow bands of frequencies characteristic of stellar elements such as hydrogen and helium, much as radio stations have their own distinctive transmission frequencies. These characteristic "spectral lines," which occur whenever hot atoms glow, provide a kind of fingerprint for each chemical element.

As astronomers examine the spectra of distant stars, they find a curious phenomenon. In most galaxies the characteristic spectral lines for hydrogen and other elements are shifted toward the red part of the spectrum, toward lower frequencies. These redshifts reveal that galaxies are moving away from us.

The First Evidence of Missing Mass

In the 1930s, Swiss astronomer Fritz Zwicky measured redshifts of several individual galaxies in the Coma Cluster, an immense clump of thousands of galaxies more than 100 million light-years away. As might be expected from a swirling cloud of galaxies, he found that these bodies are moving at different speeds. The problem is that they're moving much too fast *relative to each other*. The visible masses of individual galaxies, estimated in the quick and easy way from the number and average mass of visible stars, appear to be too low to keep the cluster intact. Given the weak resultant gravitational force among the galaxies, and their high relative speeds, the cluster should quickly fly apart. Zwicky found that there isn't enough visible mass to hold the cluster together.

Two possible explanations occurred to Zwicky. Perhaps, he thought, galactic clusters are short-lived phenomena and they do, indeed, disperse. But clusters of galaxies are so common that they appear to be a stable feature of the universe, not some chance, transient event.

The only other explanation, Zwicky maintained, was "missing mass." For some reason, these galaxies are much more massive than they appear to be—they are held together by some invisible gravitational glue. For many years Zwicky's observations remained an unsettling curiosity. As so often happens in science, better instruments were needed to resolve the mystery.

The enigma of the missing mass remained unexplained until the 1970s, when improved instrumentation allowed astronomers to measure redshifts from specific regions of a galaxy. Astronomer Vera Rubin didn't expect to make history when she methodically studied the motions of stars in galaxies. She had long been puzzled by the diverse sizes and shapes of galaxies, which come in football shapes, elegant spirals, spheres, and irregular blotches. Perhaps, she thought, by studying how these vast collections of stars move, an explanation for the variety of galactic shapes would emerge.

At one level, Rubin's task was straightforward, though so exacting it required the most advanced telescope hardware available at the time. All she had to do was obtain redshift data from different parts of a variety of galaxies. Spiral galaxies, which feature a swollen, bright central region and stately curving arms, are expected to rotate in a particularly straightforward way. These common galaxies thus provided an important test of Rubin's method, which involved measuring redshifts across the entire galaxy, from one limb to the other. As a spiral galaxy rotates, one limb moves toward us while the other moves away.

Rubin's work demanded long, lonely night-time hours collecting data with the four-meter telescope at Kitt Peak in Arizona. Pushing the telescope to its limits, she focused on the luminous disks of more than a dozen relatively close galaxies that appear slightly tilted to Earth. Each redshift spectrum required two to three hours of exposure time; four spectra of four different galaxies per night was the most she could hope to measure.

After four tedious nights measuring spectra, a strange and unexpected pattern began to emerge. Rubin had expected that rotation speeds would fall off near the outer edges of each galaxy, where mass should be least concentrated and gravitational forces should be weakest. No such trend appeared. Instead, stars at the extreme limbs of spiral galaxies orbited at virtually the same speed as those closer in. How could that be? Based on the stars we see concentrated in the very massive central bulge, central stars should orbit rapidly while matter near the less dense galactic edge should orbit much more slowly.

Rubin was amazed to discover that outer portions of spiral galaxies rotate two to three times faster than they should based on the gravity produced by stars we can see. Her original question about the different shapes of galaxies was temporarily set aside as she puzzled over her remarkable findings. Thinking about the way stars orbit a galaxy, she realized that the simple equation describing orbits has only three variables: orbital distance, orbital speed, and galactic mass. Rubin had carefully measured distance and speed, so she had to conclude that estimates of mass based on visible stars must be wrong; there must be a great deal more mass than what was visible. Most of the matter in the universe, she claimed, is dark and invisible.

How Is the Missing Mass Distributed?

Photographs of galaxies reveal well-defined boundaries. Stars concentrate in the sharply outlined central bulge and spiral galactic arms. Indeed, it seems an easy task to measure the size and shape of most galaxies from these photographs, and it might seem logical to assume that the missing mass is concentrated in this luminous region. But studies of the past two decades reveal that galaxies aren't what they first appear to be. The star-studded portion of most galaxies, like the tip of an iceberg, is often only a small fraction of the entire structure. Most of the dark matter may be concentrated in a vast "halo," many times larger than the visible region.

This conclusion was reached in the late 1970s, about the time Vera Rubin was discovering the unexpectedly rapid rotation of the limbs of spiral galaxies, and it has been amply supported by subsequent observations. Radio astronomers, focusing their giant metal dishes on the periphery of nearby spiral galaxies, found evidence that hydrogen atoms form a tenuous cloud extending hundreds of thousands of light-years *beyond* the visible portions of some of these bodies. Delicate wisps of hydrogen atoms can be detected because they emit characteristic radio waves at a frequency of 1.5 billion cycles per second, just slightly higher than the frequencies used by UHF television stations. Radio astronomers determine the velocities of hydrogen clouds by measuring the redshift of this distinctive hydrogen radio signal.

Remarkably, these swirling clouds of hydrogen atoms orbit the galaxy just as fast as the visible stars much closer in. The conclusion: Many galaxies have a massive invisible halo extending perhaps a dozen times farther than the inner visible region—a halo that may hold more than 90 percent of the mass of some spiral galaxies. Based on the weakness of the hydrogen signals, astronomers estimate that the total mass of the hydrogen atoms themselves is trivial, and not nearly enough to explain the missing mass. The motion of the insubstantial clouds indicates that something else makes up the halo of dark matter, much as the movement of puffy white clouds on a breezy day reveals the behavior of the invisible wind.

Dark Matter in the Milky Way

The behavior of distant galaxies is one thing, but what of our home, the Milky Way spiral galaxy. Does it, too, contain hidden matter? That question has been answered recently with a resounding yes by redshift studies of stars and hydrogen gas at the outskirts of the Milky Way. Based on the distribution of the Milky Way's visible mass, stars shining near the visible perimeter of our galaxy should move in orbits at a leisurely pace of about 100 miles per second. But redshift measurements confirm that stars in the outermost portions are moving at an average velocity of more than 200 miles per second. Our galaxy, like most others, must contain immense quantities of unseen matter.

What is this mysterious missing mass that may form 99 percent of our universe. How can we describe something that we can't see, touch, taste, or smell? Where is this missing mass that none of our instruments can detect? What is the nature of this invisible matter? Scientists focus on two possibilities, one as mundane as rock, the other as exotic as anything we can imagine. Many scientists now conclude that both answers may be correct.

WHAT IS DARK MATTER MADE OF?

The simplest explanation to a puzzle is often the best. Virtually all the described mass in the universe is in the form of atoms and the particles that comprise them—material given the imposing name "baryonic matter" by physicists. Before looking for strange and unknown forms of dark matter, astronomers first must inventory all the ordinary baryonic objects in the sky that don't glow. In their continuing search for amusing acronyms, astrophysicists have given the most likely candidates the collective name, MACHOs, for massive compact halo objects.

The lengthy list of MACHOs includes the most familiar bodies in our solar system: planets, moons, asteroids, and comets. In the solar system we see these bodies when they reflect the sun's light, but drifting in the depths of interstellar space they would be dark and undetectable by our most advanced telescopes. Similar bodies must litter the universe, but no one knows how many.

Also included on the list of MACHOs are a host of large, warm objects that emit infrared radiation (what we feel as heat). Brown dwarfs are planetlike objects much larger than Jupiter, the largest planet in the solar system, but too small to be a star. Brown dwarfs would be warm with heat left over from their formation, but they wouldn't visibly glow. Though difficult to see from Earth, the first convincing evidence for such a body was detected by the Hubble space telescope late in 1995. Perhaps some of the missing mass is nothing more than these big clumps of hydrogen and helium that never got large enough to be stars.

Another source of some dark matter may be ancient stars that have burned up their nuclear fuel and cooled to a dull red glow, rich in infrared radiation, but much too faint for us to see as visible light from Earth. Are the halo regions of galaxies strewn with these massive burned-out hulks of stars? Preliminary surveys from infrared telescopes recently placed in orbit around the Earth have failed to locate enough warm objects to account for more than a small fraction of the missing mass, though the search continues.

Other candidates for baryonic dark matter fall at the opposite extremes of mass. On a small scale, tiny snowballs of hydrogen ice could represent some of the missing mass. Such small objects, even in great abundance, would be almost impossible to detect from Earth. At the largest scale of mass are black holes—objects so massive that not even light can escape from them. Recent observations indicate that giant black holes, some with masses equal to millions of suns, lie at the center of many galaxies. If such massive black holes are common, then the universe may contain much more mass than is evident from starlight. Though black holes may seem by definition unseeable, they aren't actually black. They can light up space by sucking mass from nearby stars in spectacular jets of glowing gas.

Other black holes reveal themselves by bending the light from more distant galaxies. In his general theory of relativity, Albert Einstein predicted that massive objects bend light, and his prediction has been borne out by many observations of gravitational lenses. Intervening, invisible massive objects such as black holes distort the images of more distant light sources. As a result, remote galaxies may appear warped and bent by such lensing effects, like objects viewed at the bottom of a swimming pool.

Closer to home, a similar phenomenon works on a much smaller scale. MACHOs about the size of Jupiter or larger can behave as miniature gravitational lenses. As the Earth orbits the sun, a MACHO in the halo of the Milky Way would appear to move back and forth against the more distant backdrop of stars in nearby galaxies. If the MACHO appears to pass near a distant star, the lens effect will cause the star's brightness first to increase and then decrease over a period of days or weeks. One way to determine the number of MACHOs in any region of the sky is to measure the frequency of these distinctive microlensing events. Watching any one star, the chances are slim that such an event will be seen, but by monitoring millions of stars at a time, microlensing events may appear with regularity.

How do we do this? Before the age of supercomputers, such a task would have been unthinkable, but today images of ten million stars can be compared automatically in the search for microlenses. The first results from this long-term exhaustive search are already in. For four hundred nights California astronomer Charles Alcock and coworkers used an Australian telescope to monitor 8.6 million stars in the Large Magellanic Cloud, one of the closest galaxies to our own. They predicted that fifteen microlensing events would be observed if MACHOs account for all the missing mass. In fact, during more than a year of searching they found only three MACHO microlensing events. Their conclusion: MACHOs represent only about 20 percent of all the suspected dark matter in the halo of the Milky Way.

Alcock's results could be misleading. Perhaps MACHOs are distributed unevenly and the Large Magellanic Cloud lies in a direction that is sparsely populated. On the other hand, it may be that dark matter consists of something entirely different.

Hot Dark Matter

If MACHOs account for only 20 percent of the missing mass, there must be other candidates. For the past decade theoretical physicists have focused on the possible existence of two groups of exotic, nonbaryonic particles—"hot dark matter," composed of particles that travel near the speed of light, and "cold dark matter," which includes a potpourri of more sedate objects.

It's not easy to describe a part of nature that we can't examine and measure. These shy objects can't have an electric charge, for example, because charged particles like protons and electrons exert large forces on the world around them. They can't interact with light or other forms of electromagnetic radiation; otherwise we could see them. They must be very small, because they pass right through the largest objects in the universe without any discernible interaction. And they must have significant mass, comparable to the nucleus of an atom, or else they must be present in inconceivably large numbers, to account for gravitational effects of the missing mass.

Neutrinos are at present the leading hot dark matter candidate, and thus a promising source for much of the universe's missing mass. Neutrinos are well-known energetic particles because they carry energy away from many nuclear reactions and they are produced in abundance by the sun and other stars. Yet even though they have been studied for decades, it has proven extremely difficult to detect neutrino mass, which may be only about a millionth that of an electron.

Neutrinos come in three distinct varieties, or what physicists call flavors, and each experiment to detect neutrinos is sensitive to only one type. The key to detecting neutrino mass has been to look for oscillations—spontaneous changes in neutrino flavor during an experiment. Because of a quirk in particle behavior, such changes can occur only if two flavors differ in mass: Hence, if oscillations are detected, then at least one flavor of neutrino has mass.

The first tantalizing evidence for neutrino oscillation came from studies of the sun, which seems to radiate only about a third of the neutrinos predicted by theory. Astrophysicists speculate that during their eight-minute trip to Earth, two-thirds of solar neutrinos may oscillate to flavors that cannot be detected, proving that some neutrinos have mass. Of course, it's also possible that theories of solar neutrino production are in error.

The most suggestive experiments, which are now taking place at Los Alamos National Laboratory's Isabelle particle accelerator, involve producing large numbers of neutrinos and measuring how they change as they fly twenty-nine meters to a massive detector. Preliminary results are compelling. Over a four-month observation period in 1994, oscillations were detected in eighty neutrinos, many more than could be explained by experimental error. These results will have to be duplicated at other facilities, but many scientists hope that the question of neutrino mass may soon be solved.

In contrast to neutrinos, axions are a long-shot candidate for the missing mass. They are the purely hypothetical invention of physicist Frank Wilczek, who needed such a particle in his theory of the stability of atomic nuclei. If axions exist, they are predicted to have hardly any mass at all—less than a quadrillionth of the mass of one hydrogen atom. But this lack of mass could be offset by their sheer abundance—if they exist.

WIMPs

The search for missing mass is ample justification for speculating on exotic particles, but astrophysicists who strive to understand the earliest history of the universe have additional reasons for wanting to find dark matter. They need such matter to explain one of the most puzzling aspects of the large structure of the universe—its lumpiness.

Based on what we know about atoms and their building blocks, the events following the big bang origin of the universe should have flung matter outward in a uniform mix. In the unimaginably hot and bright first moments of the universe, extreme temperatures and the radiant pressure of light would have prevented ordinary atomic building blocks like protons and neutrons from clumping together. By the time the universe cooled down enough to form ordinary atoms, it would have been too dispersed to develop large objects like galaxies that are bound together by gravity. The trouble with this scenario is that we *do* see clumps: galaxies, clusters of galaxies, and huge superclusters of galaxies, separated by gaping voids more than a hundred million light-years across. How could theory be so wrong?

Perhaps, physicists surmised, the answer lies in the second kind of exotic, nonbaryonic particles—slow-moving "cold dark matter," or WIMPs, for weakly interacting massive particles. These particles don't interact with light, and so they wouldn't have been pushed around by the extreme concentration of radiant energy in the early universe. Cold dark matter could have concentrated in galaxy-sized objects while the universe was small, long before atoms were cool enough to form stars. (Even if neutrinos are found to have mass, their properties cannot explain the clumping of the early universe because neutrinos travel too fast to have formed clumps in the early universe.) Thus, for physicists hoping to explain how the universe's large-scale structure arose, WIMPs provide an elegant solution.

WIMPs, particles of cold dark matter with predicted masses about ten times that of a hydrogen atom, are thought by many scientists to be the most likely missing-mass candidates. These hypothetical particles seem especially promising, not only because they fulfill all the requirements for the missing mass, but also because they were proposed to exist quite independently of the dark matter problem by physicists attempting to devise a unified theory of natural forces.

Cold dark matter continues to provide scientists with a chance to speculate on the nature of the unseen universe. Among the more exotic suggestions are "macroscopic quark nuggets," which are essentially gigantic nuclear particles; magnetic monopoles, which possess only a north or a south magnetic pole; and miniature primordial black holes formed at the time of the big bang. Physicists also contemplate supermassive "cosmic strings" and "cosmic loops," structures formed in the earliest instant of the universe that may have once stretched across the entire universe. Ultimately, however, these scientific ideas are useless unless they make specific, testable predictions. That is why more than a dozen groups of scientists around the world are actively searching for one of nature's most elusive prizes.

Searching for Something
That Might Not Be There

The key to discovering WIMPs and other forms of nonbaryonic dark matter lies in the hope that they are *weakly* interacting particles. If they are to be found, every so often one of these tiny objects must jostle an atom. If dark matter never interacts with ordinary matter, then it cannot be studied except indirectly by its pervasive and poorly constrained gravitational influence on other bodies. In that case, dark matter studies will have to be abandoned to philosophers and poets: If researchers can't observe or measure a phenomenon, then it lies beyond the domain of science. Scientists, therefore, adopt an optimistic assumption in their quest for this elusive missing mass. They must assume that these camera-shy particles occasionally interact with atoms in the Earth to produce a distinctive, if minuscule, burst of energy—energy that can be detected by a well-designed experiment.

Each of the prime dark matter candidates—neutrinos, axions, and WIMPs—is predicted to interact with matter in a distinctive way. Neutrinos are the easiest of the three to detect, and they are actively studied in a dozen labs around the world. They whiz by the millions through every square inch of the Earth every second. Almost all of them pass by untouched, but a few will hit an atomic nucleus, causing the atom to vibrate. These atomic vibrations, in turn, induce the tiniest electrical current, which can be measured by supersensitive detectors.

The search for neutrinos takes place deep underground, in mines and caverns that shield sensitive instruments from the incessant bombardment of cosmic rays. The greatest enemy of all seekers of nonbaryonic dark matter is "noise"—random unwanted signals from electronics, ground vibrations, and background radiation that pervade every part of the Earth. Some neutrino experiments, such as the one at Los Alamos, rely on huge tanks of chlorine-rich liquid that is especially susceptible to light-producing neutrino interactions. Other neutrino detectors are crafted from ultrapure germanium and silicon crystals, cooled to near absolute zero, and surrounded by a wall of lead shielding.

Axions are much more elusive than neutrinos, and must be captured by more extreme means. Individual axions, if they exist, simply do not have enough energy to interact with atoms. The only hope for axion detection was proposed by Belgian theorist Pierre Sikivie, who realized that an axion passing through an intense magnetic field might be disturbed just enough to produce a weak microwave signal. Teams of physicists in New York and Florida have spent several years, searching without success, in attempts to create and amplify this microwave signal. Axions are so elusive, however, that many years of observations and significant improvements in instrument sensitivity may be required before a single definitive signal is found.

According to some particle physicists, WIMPs hold the greatest promise of supplying the missing mass, but they also represent the greatest challenge in detection. Preliminary, unsuccessful efforts to capture WIMPs employ underground germanium detectors similar to those that record the existence of neutrinos. The formidable difficulty in WIMP detection is that their collisions with atoms may produce only a thousandth of the energy observed in neutrino collisions—far too weak a signal to observe above the incessant noise of current hardware. The principal consequence of a WIMP–atom interaction will be a minuscule increase in temperature, perhaps less than a millionth of a degree. Researchers are now developing a new generation of particle detectors, including radically new designs employing superconducting electronics, to continue the search for these invisible forms of matter.

By the most optimistic estimates, if researchers are resourceful and lucky, the mystery of cold dark matter might be solved within a decade or two.

What Will It Mean to Find a WIMP?

Great scientific discoveries change the way we think about ourselves. Astronomers have proven that the Earth, once believed to be the center of the universe, is a relatively small planet orbiting one of a hundred billion stars, in one of a hundred billion galaxies. Now the discovery of missing mass suggests that atoms, the stuff of which we are made and the only matter we know, may constitute only a tiny fraction of what exists out there.

For all of human history we have known only one kind of matter—atoms and the particles that make them. How astounding it is to think that most of the universe's mass might be something else. We are confronted with so many surprising mysteries: What is this strange stuff? How can we study it? What laws govern its behavior? And if we can confine and shape this matter to our will, what undreamed of technologies might follow?

Fate: Will the Universe End?

I don't know why we should care whether or not living organisms
will still exist billions of years from now, but for some odd reason
most of us apparently do. We are disturbed by the fact that the
earth is destined to be vaporized when our sun becomes a red
giant in about 5 billion years, and we find it depressing to
contemplate that the universe will eventually become lifeless.

—Richard Morris, *Cosmic Questions* (1993)

Some say the world will end in fire,
Some say in ice.

—Robert Frost, *Fire and Ice*

The Family Photo Album of the Universe

If you look through a favorite family photo album, you'll probably find
pictures of festive weddings, newborn babies, and children's birthday
parties—rites of passage in all our lives. You might pause at the image
of a favorite grandparent who died recently, or your first child just
learning to walk. Dates and places written beneath each picture put
events in sequence. A photo album provides a cross section of many
people's lives.

In fact, the human life cycle can be reconstructed from an album of
separate snapshots. Each image, with its accompanying data, provides
clues about a specific place and moment in time. By examining lots of
photos, and placing them in some kind of order, it's possible to deduce
aspects of how humans grow up and grow old, as well as gain insights
about family structures, social rituals, and human behavior. Given
enough photographs, a careful outsider could develop theories that answer a lot of questions about the human species.

That's very much how cosmologists attempt to discover the evolution and ultimate fate of the universe. Looking into the heavens, we see
snapshots of the past. From the sun, which we always see as it was eight
minutes earlier, to the most distant galaxy, shining with light emitted
many billions of years before the Earth was born, we are always looking
at snapshots of the past. We can never know the state of the universe
right now. Nevertheless, with the most powerful telescopes, we catch
glimpses of the extreme limits of visible matter. With enough data on

these distant sources of light, we begin to deduce details of the ancient beginnings of the universe. We can also, or so cosmologists hope, guess its ultimate fate.

Cosmic Beliefs

We are locked into our own spacetime. Light, rushing through space at 186,000 miles per second, carries our only source of information about the far reaches of the universe. We observe only those events in time and space that coincide with the passage of this ancient light.

By studying these snapshots of the universe, we can measure spectra of stars and galaxies, learn of their temperatures and compositions, and infer details of their past and future evolution. Even so, much about the universe remains intrinsically unknowable. As a result, the study of the universe as it was, is, and shall be is a science rich in philosophy and belief. Every model of the universe incorporates assumptions about how the cosmos ought to be that can never be verified. As physicist and philosopher Edward Milne wrote: "One cannot study cosmology without having a religious attitude to the universe. Cosmology assumes the rationality of the universe, but can give no reason for it short of a creator of the laws of nature being a rational creator."

Principal among the articles of cosmic faith are assumptions about whether the universe is *homogeneous*, *finite*, and *static*.

Is the Universe Homogeneous?

One of the most widely held cosmological beliefs is an assumption that the universe must appear essentially the same to every observer, everywhere, at any given time. According to this assumption, the only distinctive portion of the universe for any observer is his or her immediate vicinity, with its idiosyncratic array of planets and stars. At larger scales, the distribution of objects tends to average out. This is not to say that matter is smoothly distributed throughout the universe; clusters of galaxies are anything but uniform. But every observer sees essentially the same degree of nonuniformity. In the words of cosmologist Edward Harrison, "Variety in thought and outlook is to be found within a galaxy and not by exploring the utmost depths of the universe." This remarkable, sweeping concept, championed by Albert Einstein and many others, reflects a rather recent point of view—one that was not held by geocentric philosophers of past centuries.

Observational data cannot prove homogeneity. From our vantage on Earth, we find no evidence that there is anything special about our home. When averaged out to the scale of galaxies, mass and energy appear to be isotropic—more or less equally distributed in all directions. Galaxies are not more common in the northern than the southern hemisphere, for example. But isotropy does not necessarily prove homogeneity. After all, our galaxy could be at the very center of a spherical universe and thus the heavens might appear the same in every direction, like the view from atop a solitary mountain peak. As we peer farther and farther from Earth, we can look only into the symmetrical past. We have no way of knowing if the rest of the universe, right now, is like our neck of the woods. It requires a leap of faith to assume that every place is the same just because every direction looks the same. Nevertheless, that assumption of homogeneity is central to all current cosmological models.

Is the Universe Infinite?

Astronomers must resign themselves to the nagging fact that most of the objects in the universe can be observed only as they were in the distant past. Even more troubling, however, is the realization that we can study only the *visible* universe. The size of the actual universe may be vastly greater than what can be seen. If some objects are so far away that the intervening space is expanding faster than the speed of light, then we cannot possibly know of their existence, for their light can never reach us through the intervening distance.

Scientists, unable to prove how large the cosmos really is, create models that reflect how big they think it should be. Most cosmologists believe in a universe of finite extent: An infinite universe presents what many philosophers consider to be unacceptable paradoxes. An infinite universe holds every possible combination of atoms *an infinite number of times*. The human body contains about a hundred trillion cells, each with a hundred trillion atoms. The total number of different combinations of those trillions of trillions of atoms is vast beyond imagining, yet that huge number is nothing compared to infinity. Every conceivable combination of atoms must repeat over and over again in a truly infinite universe: An infinite number of you has sat in a chair like your chair, reading this book. Nothing is unique in an infinite universe. Scientists cannot disprove the infinite universe model, but most philosophers balk at such a concept.

Is the Universe Static?

Albert Einstein, following the publication of his general theory of relativity, postulated a static universe, one eternally fixed in size. Spacetime in his vision was permanently curved and warped by gravity. To maintain his static universe, Einstein had to introduce a hitherto unknown cosmic force, described by a "cosmological constant," that offset what would otherwise be an inevitable gravitational collapse. At first he believed his constant to be an indispensable aspect of cosmology, but, following the discovery that the universe is expanding (see below), he abandoned the notion and regretted introducing a needless complication. Thereafter, he called the cosmological constant "the biggest blunder" he ever made in his life. Many other scientists, however, continue to speculate on the utility of this convenient constant, which can shape the cosmos to almost any desired past and future by offsetting or modifying gravitational effects over the vast distances of expanding space. For now, at least, the introduction of this "fifth force" into models of the universe is a philosophical choice rather than an observational imperative.

The Expanding Universe

Astronomer Edwin Hubble helped to discover two remarkable facts: The universe is immense and it's getting bigger. These profound insights were made possible, in part, by the newly commissioned 100-inch telescope of the Carnegie Institution of Washington's Mount Wilson Observatory in the 1920s. Hubble trained the giant apparatus on remote galaxies in an effort to answer a long-standing question about the distance to these fuzzy-looking objects.

Before Hubble's work at Mount Wilson, astronomers weren't sure if galaxies were nearby clouds of gaseous material in the neighborhood of our own Milky Way, or if they were much more distant collections of stars. Hubble's beautiful galactic photographs resolved individual stars in many distant galaxies, thus proving that they lie far outside the Milky Way. He also found that immense numbers of previously unobserved galaxies litter every part of the skies, thus increasing the size and mass of the known universe by many orders of magnitude.

Hubble's discovery of the nature of galaxies introduced a new era of cosmological research, but his most extraordinary contribution came when his estimates of galactic distances were combined with measurements of the spectra of these galaxies. Early in this century Vesto Slipher in Arizona and Willem de Sitter in Holland had found that distant galaxies have significant redshifts—the galaxies appear to be moving away from us at high velocities. Hubble's data revealed that the more distant galaxies were receding more quickly; a galaxy twice as far away appears to recede twice as fast. This pattern is so distinctive and so uniform that Hubble and his contemporaries had to accept an astonishing conclusion: The universe must be expanding.

Stretching the Fabric of Space

The concept of expansion is easily misinterpreted. Distant galaxies are not speeding away from us through fixed space as they might after an epic explosion that began at a specific location in space and blasted matter out through that space. Rather, Hubble's concept of expansion applies to all of space itself. Distant redshifted galaxies are essentially stationary objects in a uniformly expanding universe. The galactic redshifts that we observe, in sharp contrast to the rotational redshifts observed by Vera Rubin in dark matter studies, are not the result of a Doppler effect, but rather reflect the literal stretching out of light waves by the expansion of space in which they travel. If, in a given time period, the size of the universe doubles, then all wavelengths of light traversing the universe during that same time period must also double, and will appear to be redshifted a corresponding amount.

The consequences of universal expansion are astonishing. Solar systems, star clusters, galaxies, and even clusters of galaxies can remain gravitationally bound to one another and do not, themselves, expand.

The space between these relatively dense clumps of matter, however, increases constantly.

One surprising implication of universal expansion is that some objects, though stationary in their own space, may be receding from us faster than the speed of light. If we measure visible objects five billion light-years away receding at half the speed of light, then objects ten billion light-years away must be receding at light speed. We must, therefore, accept the consequence that the expanding universe may hold objects that we can never observe. Light rays from those objects can never reach us, and vast portions of the heavens may remain forever unknown. Like a sailor at sea, we cannot see beyond our horizon. Light emitted from objects beyond the horizon, receding at light speed, will never reach us unless the universal expansion eventually slows down or reverses.

The Big Bang

By the 1930s, observations of galactic redshifts had convinced virtually all astronomers that the universe is expanding, but expansion does not, by itself, imply a beginning of time. Not until the 1960s did observations place dramatic constraints on expansion models.

As cosmologists grappled with the strange consequences of expansion, they divided into two camps: the "bangers" and the "antibangers." With no clear evidence to support either position, the lines were drawn on philosophical grounds. Some of the bangers, including the influential cosmologist and Catholic priest Georges Lemaître, believed in a specific creation event on religious grounds, and embraced any evidence that supported such a position. Lemaître's theological leanings were thinly veiled when he wrote: "These considerations, besides providing a natural beginning, supply what can be called an inaccessible beginning."

Antibangers disagreed, and thought an eternal, unchanging universe far more elegant and pleasing. Arthur Eddington, a pioneering British cosmologist (and avowed atheist), wrote in 1931: "Philosophically, the notion of a beginning to the present order of Nature is repugnant to me." Eddington and other antibangers favored models of a steady state universe, in which new mass is constantly created out of nothing throughout all of space to form new stars and new galaxies. In these models, the universe expands forever, with no evidence of a beginning point in space or time.

Steady state theories have now been abandoned. Since the mid-1960s, when Arno Penzias and Robert Wilson at Bell Laboratories first detected a steady hiss of cosmic microwave background radiation, most cosmologists have accepted the notion that the universe began in a hot, dense state at a specific point in time—the big bang. As the universe has expanded, the intense sea of primordial high-energy radiation has been stretched out and cooled to its present state, which represents a cosmic background temperature about three degrees above absolute zero. Many uncertainties remain regarding the details of the formation process, but big bang cosmology has become as close to conventional wisdom as is possible in the science of the whole universe.

The latest generation of advanced telescopes provides additional evidence that the universe is evolving. These instruments have detected mysterious distant objects called quasars, which emitted prodigious quantities of radiant energy when the universe was young. Quasars were evidently abundant early in the history of the evolving universe, but few of these strange, energetic objects are being produced now. Since the contents of the universe have changed and evolved, the universe cannot be in a steady state.

The bangers won; observational evidence always trumps philosophical opinions, though not to everyone's liking. In 1967, Oxford University cosmologist Dennis Sciama wrote: "For me the loss of the steady state theory has been a cause of great sadness. The steady state theory has a sweep and beauty that for some unaccountable reason the architect of the universe appears to have overlooked."

Why Is the Universe Lumpy?

All big bang models start with a perfectly uniform and homogeneous universe in which every place is identical to every other place. But matter is now found in clumps of galaxies, and galactic clusters known as superclusters. The origin of such enormous features remains one of the most profound puzzles in big bang cosmology, and any successful big bang model must explain the dramatic transition between the original smooth universe and the present lumpy one.

Big bang models suggest that cosmic structure first appeared when the universe was very young—by some estimates no more than a billionth of a second old. Astronomers realized that microwave background radiation might provide cosmic clues to this structure. Before the universe was about half a million years old, matter was so hot and densely packed that radiation could not travel freely through space. When the universe cooled to the point that atoms formed, light was suddenly able to move about the entire universe. The cosmic microwave background radiation we observe today survives from that moment of release. If the universe had developed any structures in the distribution of mass by half a million years, those structures should be reflected today as subtle irregularities in the temperature of microwave background radiation: Denser regions would have produced a slightly warmer background glow.

To Earth-based observers, the three-degree background radiation is remarkably uniform in every direction—to better than a few parts in 100,000—reflecting the smoothness of the early universe. But NASA's Cosmic Background Explorer (COBE), a microwave satellite launched in November 1989, held out the hope of finding extremely subtle structures in this uniform background.

The discovery by COBE of "cosmic ripples from the beginning of time," the largest and oldest structures in the universe, was trumpeted in April 1994 at a meeting of the American Physical Society. COBE project leader George Smoot of the Lawrence Berkeley Laboratory, described systematic variations in the background radiation equal to about one hundred thousandth of a degree—a tiny effect, but one of profound implications. They had discovered "fossils of creation," direct evidence of the antiquity and scale of cosmic structure.

The COBE data has been the subject of much hype by its discoverers and rapt attention by the media. "English doesn't have enough superlatives," proclaimed Smoot, who describes COBE data as "traces of the mind of God." Such exaggerated rhetoric cannot detract from the key fact: We now have hope of obtaining direct and detailed observational evidence for conditions of the early universe, and that evidence will constrain and help shape all future cosmological models.

What Happened Before the Big Bang?

Cosmologists attempt to model the history of the universe with equations. In such an approach, many properties of the universe—its size, age, temperature, and states of matter, for example—may be reflected in solutions to those equations. But the moments of ultimate creation and destruction are mathematical singularities—conditions for which equations have no solutions. When their equations reach a singularity, scientists have no answers.

Scientists cannot, any more than anyone else, understand what it means to talk of "before the universe began" or "after the universe ends." Similarly, questions about why the universe is the way it is, whether alternative universes might exist, and the nature of time presently lie more in the speculative realm of philosophy than in the concrete domain of scientific observation, experiment, or theory. We are as baffled and speechless as anyone.

How Old Is the Universe?

Two fundamental, intertwined questions lie at the heart of attempts to understand the past and predict the future of the universe: How long ago was the cosmos born, and how big has it become? Three independent lines of evidence provide estimates of the universe's minimum age: old stars, radioactive elements, and expansion rates.

Every star is a kind of cosmic clock. Theories of star formation suggest that they begin to shine rather quickly once gravity pulls in a large enough cloud of gas and dust. From that moment of first light, the star ages and evolves, consuming its finite supply of hydrogen fuel. At each stage of its life, a star contains characteristic chemicals. Astronomers, who examine the spectra and thereby the chemical makeup of many distant stars in different stages of life, use their theories to estimate stellar ages and find that many stars in ancient star clusters appear to be about 12 to 16 billion years old. Obviously, the universe cannot be younger than these oldest stars.

Radioactive isotopes provide a second type of universal clock. Large stars burn rapidly and end their frantic lives in violent supernovas that litter the galaxy with all the chemical elements. Included in that rich mixture are long-lived radioactive isotopes, including uranium-238, thorium-232, and several others. Once formed, these isotopes begin to decay at well-known rates. Present isotopic abundances point to a time of formation roughly 11 billion years ago. Because the massive stars that formed the first isotopes are believed to have exploded early in history, the universe can't be much older than 12 billion years.

Expansion rates can provide a third, independent cosmic clock. If we play the mental image of universal expansion in reverse, then in the distant past, perhaps 10 or 15 billion years ago, the universe was small and unimaginably dense. One way of estimating the age of the universe, therefore, is to measure the present rate of expansion and work backward. The problem with this method is that the expansion rate can't be observed directly; instead, astronomers must measure both an object's recession velocity *and* its apparent distance.

The expansion rate of the universe, known as the Hubble constant, is calculated as an object's velocity divided by its distance. If the universe expands uniformly, then every distant object should display the same Hubble constant. In principle, therefore, one really needs to measure only the velocity and distance to a single galaxy. (It has to be a very distant galaxy so that recession is much greater than any local motion through space.) Velocities are easily determined from redshift data, but measuring intergalactic distance is proving to be one of the knottiest problems in astronomy.

If all the objects in the universe were stationary, or at least moving slowly relative to light speed, the question of cosmic size would be conceptually simple. It's easy, for example, to measure the distance to the sun, planets, and nearby stars just by using simple navigational techniques of triangulation. But redshift data reveal that space itself is expanding, giving distant galaxies and quasars relative motions that represents a significant fraction of light speed. Light from those objects may have been traveling toward us for billions of years, *and* in a constantly expanding universe. This situation leads to a jumble of space and time, and confuses discussions of cosmic distances.

Imagine a bunch of light waves leaving a galaxy five billion years ago and beginning a journey through space at 186,000 miles per second toward what would become Earth's position. Today, we detect those light waves as the image of a distant galaxy, but much has happened since those waves began their journey. For one thing, when the light waves left the galaxy, it was much closer than it now appears. For five billion years the intervening space has been expanding as the waves sped toward us. By the same token, the galaxy is now much more distant than it was when the light left, because it, too, has been carried along by the universal expansion for billions of years.

Astronomers must distinguish among three very different distances associated with any redshifted object. The "apparent distance" is usually based on the brightness of the object we see. Astronomers identify "standard candle" stars that emit a known amount of light. In an expanding universe, the apparent distance is generally longer than the "emission distance," estimated for the time when the light started its journey, and is shorter than the "reception distance," the actual distance to the object right now.

For any given object, the observed redshift reveals how much the universe has expanded *since those light rays were emitted*. We can't say for sure if expansion has been or continues to be smooth or jerky, or if it is the same everywhere at any given time. To calculate distance from the observed expansions, we must propose a model, and calculated distances depend to a large extent on the assumptions of that model.

Popular cosmology articles and books often sidestep this tricky distance issue, and usually report only the calculated reception distance. Trouble arises because that distance strongly depends on the value assumed for the Hubble constant, which may vary by a factor of two, depending on which expert is quoted. We can observe the universe only as a series of snapshots in space and time and, as with any mixed-up collection of photographs, space and time can become rather confused.

The Search for Standard Candles

We can determine the apparent distance to a source of light if we know how much light it produces, and measure how bright the source appears. Astronomers, seeking a reproducible method to determine apparent distances to other galaxies, are attempting to identify standard candles—stars that produce a predictable amount of light.

Cepheid variable stars, which were used by Edwin Hubble in his original galactic distance measurements, remain the most reliable type of standard candle. These brilliant stars become brighter and dimmer over a period of days or weeks; the exact period between the times of maximum brightness is closely tied to the star's average brightness. The most reliable current distance scale was worked out by calibrating the brightness of variable stars that lie within a few thousand light-years of Earth, where simple triangulation can be used to measure distance. Cepheid variables are then used to determine distances for relatively close galaxies in which individual stars can be identified. The unprecedented clarity and resolution of the Hubble space telescope helps, and will allow careful measurements of distances to galaxies perhaps 100 million light-years away.

For more distant galaxies, other brighter standard candles must be used. Star clusters, planetary nebulae, and even the total brightness of spiral galaxies, all calibrated against the Cepheid variable distance scale, have been proposed as options for measuring apparent distances to very remote objects. Brief, brilliant supernovas of a common type (called 1A) are receiving special attention these days. Any star larger than about 1.4 times the mass of our sun will end its life in a supernova explosion. A type 1A supernova occurs when a star slightly smaller than this critical mass gravitationally sucks up matter from a neighbor until it just reaches supernova size. These stars explode with a well-defined brightness—a "standard bomb" more than 100,000 times brighter than a Cepheid variable, and thus visible from hundreds of times farther away. Recent observations of several of these brief events are helping astronomers to refine the cosmic distance scale.

As astronomers, measuring ever greater distances, bootstrap themselves from one distance scale to the next—triangulation, to Cepheid variables, to supernovas, and so on—the distance measurements become increasingly suspect. For the most distant objects, which would define the Hubble constant most accurately, our distance estimates are the least precise.

Ironically, no matter how carefully we observe distant lights, all these measurements upon which cosmic questions of the history and fate of the universe ultimately hinge are plagued by one maddening uncertainty—dust. If relatively little dust obscures our view of distant objects, then a faint object is very far away, implying a large, old universe that is expanding relatively slowly—a low Hubble constant. If, on the other hand, lots of intergalactic dust obscures the view, then a faint object might be much closer, suggesting a smaller, younger universe that is expanding more rapidly—a large Hubble constant. The debate rages, as universal age estimates based on these Hubble constants range from almost 20 to less than 10 billion years old.

Tremendous effort is being devoted by astronomers to resolve the dust dilemma. One promising solution, now under investigation, is to observe both infrared radiation and visible light coming from distant objects. Dust tends to absorb visible light more than the infrared, so relative intensities of visible and infrared radiation might be sensitive to the concentration of intervening dust.

Fudge Factors

Alert readers may have noticed that some recent measurements of the Hubble constant suggest an expansion rate so fast that the universe appears to be younger than its oldest stars—a seemingly impossible situation. But in modeling there is almost always a way out. Cosmologists could, for example, dust off Einstein's idea of a cosmological constant, producing a model universe with rapidly varying expansion rate. Astronomers could thus match observed distances and velocities to virtually any age.

Gravitational Lenses

Is it possible to determine distance with a single measurement? That is the hope of MIT radio astronomer Jacqueline Hewitt and her colleagues, who have proposed a novel and elegant approach to the cosmic distance problem. The method relies on radio observations of quasars, ancient remote objects that appear to be not much bigger than a star but that may emit many times the light energy of the entire Milky Way galaxy.

Occasionally, a galaxy will lie almost exactly between us and a distant quasar, causing a curious gravitational lensing effect that produces two or more distinct images of the same quasar. Each image represents a separate path of light through space, as light is bent by the gravitational force of the galaxy. What's more, each path will generally be of different length—sometimes a light-year or more different. Any variation in the quasar's radio output will be detected as a twinkle first in one image, then the second. The time delay, combined with the visual angle between the two images, can be used to calculate the immense distance in a single measurement.

Hewitt and her coworkers are now monitoring a particularly promising gravitational lens, called the Einstein Ring, in which a quasar's split image is clearly visible. The astronomical community eagerly awaits a definitive flicker that may establish the cosmic distance scale.

How Might the Universe End?

If the big bang model of the universe's origin is correct, then we can imagine only three possible scenarios for its ultimate fate: the open universe that expands forever, the closed universe that falls back in on itself, and the flat universe in which the expansion rate slows almost to zero.

The unknown factor in choosing among these three models is the density of matter in the universe, designated by the symbol Ω. Current estimates of the average "critical density" necessary for a flat universe, for which Ω is defined to be one, are about six hydrogen atoms per cubic meter, though the exact value depends on the assumed cosmic size and expansion rate. If Ω is less than one, then expansion wins and the universe is open. If Ω is greater than one, then gravity wins and the universe is closed. And if, by chance or cosmic design, Ω is exactly one, then the universe is flat. Each of these scenarios has its proponents, on both philosophical and observational grounds.

Present best estimates of ordinary matter's density hover about one hydrogen per cubic meter—only 20 percent of the mass required for a flat universe. Observational evidence thus favors an open universe. This scenario has been dubbed the "whimper" universe after T. S. Eliot's prescient line from *The Hollow Men*: "This is the way the world ends/ Not with a bang but a whimper." An infinitely expanding universe becomes ever colder and more dispersed. Eventually, after hundreds of billions of years, all the stars burn out and the universe is forever after dark and silent. Many cosmologists find such a lackluster end to life and creation meaningless and depressing—and a prospect to be shunned in devising models of the universe. In the words of physicist Paul Davies: "If there is a purpose to the universe, and it achieves that purpose, then the universe must end, for its continued existence would be gratuitous and pointless. Conversely, if the universe endures forever, it is hard to imagine that there is any ultimate purpose to the universe at all."

In the event that expansion velocity is low and matter is sufficiently dense, then the present expansion will gradually stop and reverse in a "closed" universe. In such a universe, the cosmos will eventually contract and implode in a process sometimes called "the big crunch" that mirrors the big bang. Stephen Hawking is among those who advocate such an end to time. Others note that a closed universe may attain a kind of jerky steady state—big bang, expansion, contraction, big crunch, and another big bang, over and over again.

Finally, it is possible that the force of gravity may exactly match universal expansion in an Ω = one universe. The expansion rate slows down, ever approaching but never quite reaching zero in a "flat" universe. Such a cosmic coincidence might seem improbable, but some theorists argue that it is a likely scenario based on what we see. After all, the observed density of matter in the universe suggests that there is *almost* enough stuff out there, and dark matter could make up the difference. For many cosmologists, a flat universe that could go merrily about its business for all time is the most aesthetically pleasing prospect. In fact, one of the most influential current big bang models, introduced by physicist Alan Guth of the Massachusetts Institute of Technology in 1979, incorporates an essentially flat universe. This model, which requires that the universe underwent an initial period of extremely rapid, faster-than-light expansion, followed by today's more stately rate of expansion, yields a cosmos in which expansion and gravity exactly balance each other to produce a universe of stars and galaxies. The so-called inflationary model matches so many details of the present universe that some cosmologists believe it must be correct. The major unresolved question regards the missing mass. Dark matter, which could raise Ω to the required value of one, thus has profound implications for the fate of the universe.

Observing Ω:
How Dense Was the Early Universe?

Big-bang models make exacting predictions about how matter formed, and thus contribute to estimates of the universe's density. Some astronomers are now attempting to calculate Ω based on matter left over from the first few minutes of the universe.

Any model of the early universe requires that the kinds of matter we know today gradually condensed out of the energetic maelstrom of the earliest hot universe. Within the first few seconds of the cooling and expanding universe, matter existed as protons and neutrons (relatively massive particles that form atomic nuclei), plus electrons (the much less massive particles that surround atomic nuclei). Conditions were so extreme that these particles were unable to combine to form atoms. Negatively charged electrons and positively charged protons are long-lived particles, but isolated neutrons are unstable and spontaneously decay into one electron and one proton. Thus, the ratio of protons to neutrons steadily increased until only two of every sixteen nuclear particles were neutrons, and the remaining fourteen were protons.

When the cooling universe was about 100 seconds old, neutrons and protons were able to combine to form nuclei known as deuterons, in a mix with a ratio of two deuterons for every twelve isolated protons. Pairs of deuterons then started to collide and react quickly to form helium nuclei, each with two protons and two neutrons. If the big-bang model is correct, then in a span of 200 seconds almost all nuclear particles were locked into nuclei of hydrogen (a single proton) or helium (two protons and two neutrons) in a 12-to-1 ratio, with a minor sprinkling of a few other light nuclei. The total mass of primordial helium in this scenario is thus approximately 25 percent of the universe's total mass.

The helium-forming process was not 100 percent efficient. A small fraction of deuterons—less than one for every 10,000 hydrogens—escaped the nuclear reactions and has survived to this day. Calculations show that the denser the primordial concentration of matter, the greater the chances of a collision between two deuterons and the higher the efficiency of helium production. The percentage of primordial deuteron, the survivors of the ancient maelstrom, thus provides a sensitive indicator of the density of matter in the first few minutes. The result: Estimates of hydrogen and helium formation during the big bang account for only about a tenth of the mass necessary to close the universe—a number remarkably close to the amount of visible mass.

If Ω, in spite of such an estimate, is close to one, then dark matter must make up the difference. For many scientists who accept the inflationary model, present-day deuterium concentration provides proof that dark matter must exist in abundance.

Observing Fate: Is Expansion Slowing Down?

Hopes of predicting the future of the universe, when we can know so little about its present state, may seem overly optimistic. But decades or centuries from now, when all the kinks have been ironed out of cosmic distance scales, there may be a direct way to discover our fate. Given enough accurate velocity-distance measurements, we might be able to measure the rate at which expansion is slowing down.

If expansion is constant everywhere and throughout time, then a plot of redshift versus apparent distance is a straight line—the larger the distance, the greater the redshift. But if expansion is gradually slowing down, the line will curve, because older (more distant) galaxies used to be moving away faster. Today our observations lack the accuracy necessary to demonstrate such curvature, but improved measurements may eventually provide unambiguous evidence of a changing universe.

Whatever the ultimate fate of the universe, the drama will not play itself out for many billions of years, and it will make no difference to the Earth. Our planet will have long since died. Throughout the past several billion years, the sun has gradually increased in brightness, and that trend will continue, slowly but inexorably. Over the next five billion years the sun's radiant energy will increase about threefold. By some estimates, life on Earth, which has thrived for more than three and a half billion years, may survive only another half billion years. By that time Earth's surface temperatures will have soared as a result of increased solar temperatures and a runaway greenhouse effect.

That we should ask with something more than idle curiosity about this unimaginably distant fate of the universe—that we should care about an end that in all probability no member of our own species will witness—says much about ourselves and our aspirations. Something deep inside each of us yearns to know.

Science can hypothesize about reality's future demise, but equations do little to help us deal with the implications of that seemingly inevitable total obliteration of everything that is human. Does life have meaning? Or is our existence meaningful only in the joy and comfort we bring to others of this and future generations? Perhaps one day science will be able to offer more concrete answers.

PERFECT SYMMETRY: CAN WE DEVISE A THEORY OF EVERYTHING?

There is a good chance that the study of the early universe and the requirements of mathematical consistency will lead us to a complete unified theory within the lifetime of some of us who are around today.

—Stephen W. Hawking, *A Brief History of Time* (1988)

There is no question that there is an unseen world. The problem is how far is it from Midtown and how late is it open?

—Woody Allen

Patterns: The White Picket Fence

At the center of a small New England town, two houses sit side by side, just across the street from the village square. The Cape Cod–style cottage on the right is pure white with black trim. Rows of red impatiens line the flagstone walkway to the front door, which is flanked by two identical, black-shuttered windows. A stately maple tree graces the neatly mown front lawn, and an elegant white picket fence surrounds the yard. "What a beautiful house!" passersby exclaim. "Wouldn't you love to live there?"

On the left stands a nearly identical cottage, the same size and style, but with minor imperfections. The white paint is faded and chipped, the garden is overgrown, and one black shutter hangs askew. Perhaps the most unsettling defect is the picket fence—one picket, about three feet to the left of the gate, is missing. "What a shame!" the people say and shake their heads. "Someone should fix it."

We want our world to be symmetrical—to incorporate regular and predictable patterns. Day and night, summer and winter, work and play, eat and sleep, our lives are ordered by simple patterns, and we expect no less from the universe and its physical laws.

Physicists fervently believe that the cosmos was created with elegance and symmetry, and they demand these characteristics in their models of nature. For them, a unified theory of everything, the ultimate model that explains the behavior of all known matter and forces, must possess deep and profound beauty. When we find such a unified theory, the breathtaking patterns underlying natural processes will be revealed. The theory of everything will no doubt point to gaps in our knowledge of the physical world, like missing pickets in a cosmic fence. Then the physicists will search day and night until the missing piece is found. For nature, our grand and beautiful home, could not possibly revolve around asymmetry.

The most seductive challenge in physics today is the search for a "theory of everything," a TOE—a single set of mathematical equations that describes the properties and behavior of all the different kinds of matter and forces in the universe since the beginning of time. The ambitious search for TOEs advances along parallel fronts of experiment and theory. Experimentalists, who probe the properties of matter and energy, smash atoms and analyze starlight in the hope of glimpsing the most fundamental fragments of matter. Theorists scrutinize these experimental data for hidden symmetries, applying advanced mathematical techniques to devise equations that predict reality. Perhaps, someday soon, an ultimate theory of everything will emerge. What will such a theory look like? What good will it be?

First and foremost, a theory of everything will be a cohesive set of mathematical expressions that most physicists expect to be simple and elegant. From those expressions it should be possible to derive all of the constants of nature: the speed of light, the strength of the gravitational force, the mass of a proton, and the magnitude of an electron's charge. The theory should point to the existence of the many kinds of subatomic particles that we know, and very likely to some we've never seen, such as promising candidates for dark matter. The theory should also explain the subtle relationships among the four known forces of nature: the everyday forces of gravity and electromagnetism, as well as the strong and weak forces that operate only within an atom's nucleus.

The mathematics of a successful theory of everything will, no doubt, be formidable. Current models, for example, suggest that the universe exists not in the familiar four dimensions of space and time, but in ten or more dimensions, most of which we never experience. Even if a successful TOE is found, solving the equations and understanding all their implications may take experts many decades, but the underlying concept and profound significance of such a theory should be accessible to everyone.

Implications of a Unified Theory

Few would disagree with Stephen Hawking, who states: "The ultimate theory of the universe . . . would bring to an end a long and glorious chapter in the history of humanity's intellectual struggle to understand the universe." A successful theory of everything would address one of science's greatest unanswered questions. But would it be more than that?

A small band of physicists, deeply involved in the search for a theory of everything, describe unification as *the* key unanswered question in science. Its solution, they suggest, would mark "the end of physics." Hawking pushes the hyperbole envelope, calling a successful theory of everything the "ultimate triumph of human reason—for then we would know the mind of God."

A unified theory would mark an extraordinary accomplishment, to be sure, but many scientists are uncomfortable with Hawking's theological perspective, and they certainly do not view a theory of everything as the end of an entire branch of science. Physicist Richard Feynman, grappling with this issue, placed TOEs in the larger context of religion and physics in *The Character of Physical Law*, a 1965 essay that touched on the hierarchy of ideas. One extreme of this hierarchy, Feynman reasoned, encompasses overarching laws of physics—perhaps someday including a theory of everything—that describe the universe in terms of interactions at the most fundamental level. Moving up the hierarchy of ideas, one finds ever more complex concepts: the atom, heat, a salt crystal, a thunderstorm, frogs, an ecosystem, history, beauty. "Which end is nearer to God . . . beauty and hope or the fundamental laws?" Feynman asks rhetorically. He answers, "I do not think either end is nearer to God."

It may be true that we cannot comprehend the mind of God without first understanding unifying laws of nature. But, even if we were to devise a theory that explained the workings of all forces and allowed us to predict every possible interaction of matter and energy, we would know only the ground rules of the game of the universe. Nothing in that theory could prepare us for the organizing principles of the complexity that surrounds us. Just as knowing the details of computer hardware tells us nothing about the software it runs, knowing how matter interacts tells nothing of how it is organized or how it evolves into such wondrous complexity.

THE SEARCH FOR SUBATOMIC PARTICLES

In 1869, Russian chemistry professor Dimitri Mendeleev decided to tabulate properties of the known chemical elements in preparation for a lecture at his St. Petersburg classroom. New elements were being discovered every year, and it was difficult to keep up with all the new data. As he compiled his notes on atomic weights, chemical properties, and physical characteristics, he began to notice curious trends—systematic patterns in properties that suggested a subtle organization to the dozens of entries.

Mendeleev arranged the sixty-three then-known elements in a table, with elements of increasing weight arranged left to right and top to bottom. Elements with similar properties were grouped vertically in columns in this novel scheme. Mendeleev and his contemporaries were immediately struck by two key features of this "periodic table of elements." First, the table revealed a kind of symmetry in the regular repetition of element properties. Second, obvious gaps occurred where elements were evidently missing—most notably elements 21 and 31. Within a decade, chemists had discovered scandium and gallium, the two elements predicted by holes in Mendeleev's symmetrical table.

Such is often the pattern of scientific discovery: Data are acquired, patterns observed, and new data sought based on specific predictions. And what nineteenth-century chemists did for elements, twentieth-century physicists have accomplished for the astonishing array of particles that make up atoms.

Particle physics, though almost exclusively a product of twentieth-century science, boasts a rich and complex history. British physicist J. J. Thompson discovered the first subatomic particles in 1897 when he deflected a beam of electrons in a magnetic field. A decade later Ernest Rutherford, Thompson's successor as Cavendish Professor at the University of Cambridge, bombarded a piece of gold foil with alpha radiation in experiments that led to the discovery of the atomic nucleus in 1911, and shortly thereafter the proton. Rutherford thus became the first experimenter to study pieces of atoms by watching what happens when they are blasted by other pieces of atoms.

A flurry of particle discoveries followed: the neutron in 1932, the positron also in 1932, the mu meson in 1934. Each new particle was scrutinized—its mass, charge, and interactions with other particles were measured. Almost immediately intriguing patterns were observed. The proton and neutron have almost the same mass, and together account for almost all of the mass of atoms. The electron and positron have exactly the same mass, but opposite charges. Isolated neutrons spontaneously decay into a proton and an electron. From these intriguing equivalences the first subtle hints of particle symmetry emerged.

Following World War II, many more particles were discovered as physicists examined the remnants of collisions between earthly matter and energetic cosmic rays that bombard us from outer space. Researchers had long been frustrated, however, by the haphazard arrival of their cosmic ray bullets. They needed a controlled particle beam with well-defined properties, and so began the modern era of particle accelerators.

Particle accelerators are machines that employ powerful electromagnetic fields to accelerate a beam of charged particles. Speeding particles collide, producing a shower of fragments. Massive detectors record the energetic passage of these bits, while intense magnetic fields bend particle paths to reveal their mass and electric charge.

As physicists searched for ever more massive particles, they built larger machines that accelerated particles to higher energies. High energies are required to make massive particles, according to Einstein's equation $E = mc^2$, which equates even very small mass to very large energy. Physicists measure these energies in millions (MeV), billions (GeV), or trillions (TeV) of electron volts (a proton's mass is equivalent to about one GeV). Accelerators reached energies of tens of GeV in the 1950s, hundreds of GeV in the 1970s, and a TeV in the 1980s. By the 1990s, new accelerators measured tens of miles in length and cost billions of dollars. A new European accelerator called the large hadron collider is projected to achieve energies close to ten TeV in the late 1990s.

By the late 1960s, the expectation of physicists that nature holds just a few fundamental building blocks had been displaced by an unruly and growing particle zoo. Literally hundreds of particles were found: lambdas, muons, pions, mesons, alphas, and omegas—more particles than Greek letters to name them. It was up to the theorists to restore some order to the chaos.

THE SEARCH FOR SYMMETRY

Many people think of a symmetrical object as one in which two halves exactly balance, as in a reflection. The human body possesses this distinctive and simple (and, many would argue, pleasing) bilateral left-right symmetry. Everywhere we look, nature is replete with other beautiful examples.

In the world of science and mathematics, symmetry has a broader definition: symmetry is a characteristic of any object, force, or experience in which a motif is repeated in one or more dimensions. Floral wallpaper repeats a flower motif in two spatial dimensions. A crystal repeats an atomic motif in three spatial dimensions. Rock music repeats a rhythmic motif in the dimension of time. Similarly, the symmetry of any pattern, whether in nature, the world of art, or an imaginary construct of the mind, is defined by motifs and their repetition in one or more dimensions.

The quest for a unified theory is, in large measure, a search for the symmetry of the subatomic universe. Experimentalists, using gigantic accelerators, identify the fundamental particles that constitute the motifs of the physical world. Theorists, relying on advanced mathematics and leaps of intuition, then must deduce the dimensions in which those motifs are repeated, and they must discover the "symmetry group" that defines how motifs are repeated over and over again to form the cosmos. Many physicists hold a deep conviction that such symmetry exists, and that it will be beautiful when it is found.

Symmetry guides us, but not always in obvious directions. For thousands of years astronomers were hampered by the false assumption that planetary orbits must obey the perfect symmetry of circular paths. The seventeenth-century recognition of elliptical planetary orbits allowed Newton to discover the deeper symmetry of universal gravitation, which acts equally between any two objects, anywhere in the universe. Only by abandoning the cherished symmetry of circular orbits could science advance.

Dimensions of the Cosmos

All symmetrical patterns exist in one or more dimensions. A simple motif "- -" may be repeated in one dimension along a line:

- - - - - - - - - - - - - - - -

Similarly, a two-dimensional pattern of Gs that covers a surface consists of a motif (G) and its repetition parallel to the two sides of the page:

G G G G G G G G G G
G G G G G G G G G G
G G G G G G G G G G
G G G G G G G G G G
G G G G G G G G G G

In these familiar examples, patterns occupy dimensions of space, which can be measured by a ruler.

The rules of symmetry apply to any kind of dimension that can be measured. The two-dimensional pattern of Gs, for example, could be made to flash on and off at regular intervals, producing a three-dimensional pattern in which the G motif repeats in two spatial dimensions and in the dimension of time.

With this broad definition, symmetry appears in many surprising contexts and in many subtle dimensions. We experience symmetries associated with predictable daily and yearly patterns of changing light, sound, and temperature. All forms of art rely on the repetition of recognizable patterns in sometimes subtle dimensions. In painting, dimensions of color, brightness, and transparency may be superimposed on the spatial dimensions of a canvas. Music embodies no spatial dimension, but relies instead on repeating patterns of pitch, timbre, and loudness, all of which are measurable properties of sound energy, in time. Our bodies display their own inner patterns of symmetry: Heartbeat, breathing rate, and countless daily and monthly chemical cycles are intimate dimensions of our lives.

In the world of subatomic particles, symmetries occur in dimensions completely outside our personal experience. Electrons and protons are symmetrical with respect to the dimension of electric charge: Electrons have an electrical charge of −1, while protons have a balancing charge of +1. Similarly, the electron and the positron are symmetrically opposite with respect to spin. These observations cannot be random chance— electrical charge, and spin *behave as dimensions of the universe*. We can measure these attributes of subatomic particles, and observe their repeating patterns. And while electric charge and mass bear some relationship to familiar everyday properties of matter, the equally important characteristics of "strangeness" and "color" are fanciful names for dimensions that are beyond human senses—dimensions revealed only by the striking symmetries that appear as physicists measure and catalogue the behavior of subatomic particles.

Symmetry Patterns

The human mind can imagine an infinite number of possible symmetrical patterns, but mathematicians have discovered the surprising fact that the number of possible symmetries is often quite limited. All planar patterns, including all wallpaper, brickwork, designer fabrics, and tile floors, fall into only seventeen different possible symmetry "plane groups," each of which has a characteristic way of repeating a motif. Similarly, crystallographers have found that exactly 230 "space groups" are sufficient to describe the three-dimensional repeating patterns of atoms in all known crystals. The number of possible crystal structures appears to be infinite, but every possible structure conforms to the symmetry of one of the 230 space groups.

The limited number of symmetry groups greatly simplifies the search for patterns. As correspondences appear—sets of particles with opposite charge but similar mass, for example—most symmetry groups can be eliminated.

The task facing postwar physicists searching for order in the chaos of the atom was much the same as that confronting Mendeleev as he attempted to discover order among the growing catalogue of elements a century earlier: Tabulate subatomic particles according to their similarities and differences. All subatomic particles can be divided into two groups: hadrons and leptons. Hadrons, such as protons and neutrons, occur in the nucleus and participate in nuclear reactions. Leptons, including electrons and neutrinos, exist outside the nucleus.

Murray Gell-Mann, one of many theorists who searched for patterns among the hundreds of known subatomic particles, focused his attention on hadrons, which constitute a much larger group, itself divided into baryons (like protons and neutrons) and mesons. In 1962 Gell-Mann identified a distinctive relationship among the properties of eight baryons, and a corresponding pattern for eight mesons (of which only seven were then known). This recognition led him to the largely intuitive proposition that all hadrons might be systematized in a symmetry group called SU(3), for special unitary group in three dimensions. At the time, several other theories of subatomic matter competed for acceptance, but none was so deeply rooted in symmetry, and none made such successful predictions. Gell-Mann used his model to postulate the existence and properties of several new particles—the eighth meson as well as six new baryons, most of which were found by experimentalists within a year.

In spite of this success, the discovery of order in the realm of subatomic particles was not entirely satisfying. The idea of a model with hundreds of so-called "elementary" or fundamental particles seemed wrong. Physicists suspected that another, simpler layer of reality might lie beneath all the baryons and mesons, just as the simple trio of electrons, neutrons, and protons are sufficient to form all the chemical elements of the periodic table.

THE STANDARD MODEL

Two key problems confronted physicists of the mid-1960s. First, they wanted to find a model that systematized the many known hadrons and leptons and predicted new ones. Second, they hoped to develop a model that revealed connections among the four known forces. A single theoretical structure, called the "standard model," now incorporates partial answers to both these problems.

Quarks and Leptons

By the mid-1960s a number of physicists realized that the confusion of baryons and mesons systematized in symmetry group SU(3) might not represent truly fundamental units. Murray Gell-Mann and physicist George Zweig independently observed that the complex array of particles systematized in SU(3) symmetry could be constructed from a much simpler trio of particles, just as the most complex snowflake hides the symmetry of the simple hexagon. According to this idea, protons, neutrons, and dozens of other nuclear particles are fabricated from three smaller building blocks with fractional charges of $\pm$ $\frac{1}{3}$ or $\pm$ $\frac{2}{3}$. No experimental evidence supported this novel idea and no one had ever seen one of these particles, but the idea possessed a kind of simple elegance that appealed to the physics community.

Several years of frustrating failure passed while atom smashers at Fermilab near Chicago and the European Laboratory for Particle Physics in Switzerland looked in vain for the elusive motes of matter. These experiments, which involved colliding massive protons into atomic nuclei and searching for the expected distinctive signatures in the splintered fragments, failed to yield any evidence for fractionally charged particles.

Ultimately, success came from a different kind of experiment at the Stanford linear accelerator (SLAC) in California. SLAC relies on a beam of electrons accelerated close to light speed. Unlike protons, which smash other particles to bits, electrons scatter off individual protons, much the way alpha particles scattered off gold atoms in Rutherford's original experiments that first detected the nucleus. Researchers at SLAC found that while most electrons pass right through the proton, a few are deflected to surprisingly high angles. Just as Rutherford's scattering experiment showed the inner structure of the atom, the SLAC experiment revealed that the proton itself is not uniform, but, rather, is made of several extremely tiny pieces. These objects, dubbed "quarks" by Gell-Mann, occur either in groups of three (the baryons) or as a quark linked to an antiquark (the mesons).

Gell-Mann originally predicted the existence of three different kinds of quarks, which he called up, down, and strange. The existence of all three of these was rapidly confirmed in the analysis of well-known hadrons. Discovery of new high-energy mesons in the late 1970s (for which Stanford University physicist Martin L. Perl won the 1995 Nobel prize) required two more quarks, dubbed charm and bottom. Symmetry SU(3) then required the existence of a sixth, the top quark, which was finally confirmed in 1994 after much anticipation (and a heated rivalry between accelerator physicists in Europe and the United States).

In a pleasing symmetry, exactly six leptons balance the six quarks, while six antiquarks and six antileptons complete the standard model's cast of elementary particles. The twelve quarks and leptons fall naturally into three families, each with two quarks and two leptons. The first family, containing the up and down quarks, the electron, and the neutrino, accounts for all of the familiar matter around us. Protons, for example, form from two ups and a down quark, while neutrons consist of one up and two downs. The other two families of quarks and leptons consist of high-energy particles that survive only briefly in the violent aftermath of accelerator collisions.

Universal Forces: The Gauge Bosons

Understanding the arrangement of matter's fundamental particles is only half of a theory of everything. Nature's four forces—gravity, the electromagnetic force, and the strong and weak nuclear forces—dictate the dynamic interplay of matter and energy. Therefore, we must also know how forces act between particles. Physicists expect that a single mathematical formulation will be found that encompasses all four forces, in spite of the fact that these forces behave in astonishingly different ways. Gravity pervades the universe and our lives; it holds us to the Earth, and binds superclusters of galaxies together over distances of hundreds of millions of light-years. But gravity is many orders of magnitude weaker than the other three forces. Electromagnetism, the force that causes chemical bonding and keeps us from falling through the floor, is distinctive because it can be either attractive or repulsive, unlike the other three. The strong force, which holds the nucleus together, and the weak force, which is manifest in radioactive decay, only act at the scale of an atomic nucleus.

These differences in part reflect the fact that each force operates through different attributes of particles. Gravity acts most strongly on particles with mass, while the electromagnetic force affects particles that carry an electrical charge. Strong and weak nuclear forces, on the other hand, act on particles with properties dubbed "color" and "strangeness." Each particle has its own distinctive combination of mass, charge, color, and strangeness, and so we observe widely varied behavior in the subatomic world.

In spite of these differences, physicists have developed a unified view of how forces work. According to this model, called gauge symmetry, any force between two particles is carried by gauge bosons. When two electrons approach each other, for example, the force between them is mediated by photons, the massless particle of light. Picture two skaters playing catch with a heavy ball; with each transfer of the ball the skaters recoil slightly. The weak force, similarly, is mediated by three massive particles called $W+$, $W-$, and Z_0; the strong force is carried by eight gluons; and gravity acts through an as yet unobserved gauge boson called the graviton. The standard model now posits the existence of thirteen of these gauge bosons.

Unifying the Forces

According to the standard model, 6 quarks, 6 leptons, their 12 antiparticles, and 13 gauge bosons form the fabric of our universe. All the complexity of our present surroundings evolved from what was, at the start, an exceedingly simple and symmetrical universe. If we think ourselves backward to the first moments of time, we can imagine a very different universe. Immediately following the big bang, the universe was unimaginably hot and dense. Under such an environment, matter did not exist as we know it today.

At high temperature, all manner of distinctive materials become similar. At room temperature, lumps of gold, salt, and plastic behave very differently, but raise the temperature to ten thousand degrees, and all three will blend into a uniform mixture of gases. Raise the temperature to hundreds of millions of degrees, and atoms will break apart into the more basic units of electrons, protons, and neutrons. The higher the temperature, the fewer the distinctions between different kinds of matter.

The standard model, coupled with estimates of the extreme temperatures that existed after the big bang, provides a vivid picture of the birth of the universe. Scientists now realize that discoveries in particle physics have a direct bearing on our understanding of the origins and structure of the cosmos. In the words of astrophysicist David N. Schramm, this "intellectual merger of particle physics and cosmology has been one of the scientific triumphs of the last decade." We now realize that a successful theory of everything must incorporate an exact description of the universe at the moment of creation.

Earlier than three minutes before the big bang, temperatures were so high that nuclei could not form; all matter consisted of isolated hadrons and leptons. Before the first ten millionth of a second, only quarks and leptons existed, along with the gauge bosons. Even earlier, these particles began to lose their separate identities; they were indistinguishable in a state of perfect symmetry. All the complexity we see around us is a consequence of the inevitable cooling and differentiation of an expanding universe.

Ways in Which the Standard Model Doesn't Work

The standard model enjoys great success, having effectively integrated the description of all known elementary particles and three of the four forces. Its symmetrical description of six quarks, six leptons, and their twelve antiparticles, arranged into three families, provides an elegant balance, encompassing all observed subatomic phenomena. The standard model also leads to unambiguous, testable predictions about new high-energy particles.

But the standard model is not a final theory. For one thing, the model incorporates no fewer than seventeen different numerical constants—numbers not predicted by the theory. Each of these seventeen numbers has to be measured by experimental means and plugged into the equations. A truly fundamental theory, most physicists argue, could not have so many constants. By the same token, it might be argued that a theory with thirty-seven different quarks, leptons, antiparticles, and gauge bosons must be hiding deeper simplicity.

Another difficulty with the standard model is that it fails to explain the nature of dark matter. Cosmologists still have to identify the stuff that presumably makes up most of the universe; it would be convenient if a theory of everything incorporated one or more likely candidates.

Perhaps the greatest problem with the standard model is that there is no obvious way to incorporate gravity. Unless some surprising new mathematical tricks are found, the theory cannot unify all four forces. It is just possible that such a trick has been found in the guise of super-symmetry.

SUPERSYMMETRY

Late in the afternoon on a smoggy day the sun appears as a glowing red disk low on the horizon. Under such conditions, the sun seems to be absolutely flat and two-dimensional, and nothing we can observe suggests otherwise. From our vantage point, the two-dimensional description would not be wrong; it would just be incomplete.

A cube, displayed on end, creates a similar illusion. The end view of a cube reveals a square with four corners and four lines of mirror symmetry. The square we can see, however, is just one of six faces of the three-dimensional cube. This new, higher-dimension symmetry, which incorporates all the characteristics of the original description, is called a supersymmetry. The two-dimensional symmetry of the square is completely embedded in the cube's three-dimensional symmetry.

Some of the most intriguing proposals for a unified theory begin with the standard model, and invoke supersymmetry that bumps the universe up to higher dimensions. This mathematical trick, though so far unsupported by any experimental observations, provides physicists with various paths that ease around some of the thorniest problems of the standard model.

Sparticles

The standard model posits two kinds of particles. Ordinary matter is made of fermions (the quarks and leptons), while forces are mediated by gauge bosons. These two groups display one critical difference: Fermions spin about their axes at half-integral rates of $1/2$, $3/2$, $5/2$, and so on times a fundamental rotation speed, while bosons always have integral spin rates of 0, 1, 2, and so on. One unavoidable consequence of this difference is that fermions and bosons can never change one into the other. Matter particles like electrons, and force particles like photons, are now as different as objects can be in our universe.

Perhaps it was not always so. Physicists suspect that fermions and bosons were completely interchangeable at the hot, dense beginnings of time. They became distinct entities as a result of universal expansion and cooling, at the same instant that the force of gravity separated from the other three forces, when the universe was approximately 10^{-43} second old (that's one ten million trillion trillion trillionths of a second!).

Theorists now face a dilemma. Any complete theory that unifies gravity with the other forces must also allow for transitions between fermions and bosons. But half-integral fermions of the standard model can't transform into integral spin bosons.

Supersymmetry provides an elegant mathematical solution by postulating an additional, hidden dimension to the universe. In the supersymmetric early universe, every fermion had a partner with integral spin—each quark had a corresponding "squark," for example—and these particles could transform back and forth. Similarly, every boson has a twin with half-integral spin; the photon (spin equal to one), for example, was paired with a photino, identical in every way except its spin of $1/2$.

Every known particle has a new unseen double, or "sparticle," in this strange cosmic model. Just as a cube has twice the number of corners as a square, so, too, does the supersymmetric universe have twice the number of particles as the standard model. And just as some corners of a cube are always hidden from view, so, too, are squarks and photinos. These supersymmetric sparticles may now constitute a kind of subtle and elusive parallel to the everyday world of matter and forces.

Supersymmetry theories suggest that two dramatic changes occurred in these particles when gravity became a separate force and the symmetry between fermions and bosons was lost. First, interactions between ordinary matter and their supersymmetric counterparts suddenly became extremely weak. Consequently, if such sparticles exist, they would be extraordinarily difficult to detect. Second, many of the sparticles could be extremely massive. The photino, for example, is predicted to be perhaps forty times more massive than a proton.

In other words, sparticles are elusive and massive, the essential characteristics required of dark matter.

Unsolved Problems with Supersymmetry

Supersymmetric models of the universe remain a forefront area of physics research. Several elegant variants called superstring theories, which have been explored intensively during the past decade, attempt to model mathematically all particles as tiny vibrating string segments or loops in ten-dimensional space rather than as points. We experience only the four dimensions of space and time, while the other six dimensions curl up tightly on themselves, much as a tightly rolled two-dimensional piece of paper may appear as a one-dimensional line from a distance. According to these theories, vibrational behavior of these strings accounts for particle properties such as mass and spin.

A half dozen or so different superstring models now compete; each must make testable predictions about the universe to gain acceptance. One of these intriguing models points to the existence of a shadow universe with its own complement of particles and sparticles, doubling again the number of building blocks. Could shadow matter provide more dark matter candidates?

A complementary approach, called duality, has been hotly pursued since new evidence was presented by several groups at a March 1995 conference at the University of Southern California. This model posits a kind of symmetry between fundamental particles and the larger composite objects they create. Physicists apply the term duality anytime two apparently different theories end up giving similar results. (In quantum mechanics, for example, atomic-scale objects simultaneously exhibit behavior associated with both waves and particles, or "wave-particle duality.")

In duality's bizarre mathematical vision of reality, strings like those associated with individual quarks can be described as being intricately woven from much larger composite objects called monopoles; but monopoles can also be described as being woven from quarks. Similarly, it appears that some theories of strings and black holes could be dual— both theories providing solutions for the other. As more and more of these surprising dualities emerge from theoretical calculations, many physicists suspect they are closing in on a true theory of everything.

Supersymmetric theories point to a mote of time at the very instant of creation when everything was perfectly symmetric. We are now left, however, with a pretty messy state of affairs. Can there really be 74 fundamental particles: 6 quarks, 6 leptons, their 12 antiparticles, 13 bosons, and a matching 37 sparticles?

Other difficult questions remain unanswered. What is mass? How does gravity fit in with the other forces? Can these supersymmetrical models make any *testable* predictions? Perhaps Steven Hawking is right and we are near a final theory, but it will take a tremendous leap of insight to understand what it all means.

A UNIVERSE WITHOUT CHOICES

"How much choice did God have in constructing the universe?" Einstein once asked. A few physicists suspect that the answer is almost none at all: The universe is the way it is, they say, because it couldn't be any other way.

For us to be here, to ask questions, demands a universe in exquisite balance. Were gravity slightly stronger, stars could not burn for billions of years; slightly weaker and they wouldn't burn at all. Similarly, the relative magnitudes of the strong, weak, and electromagnetic forces poise on a razor's edge to enable particles to form atoms and atoms to form molecules. Had the charge on the electron or the mass of a proton been much different, the universe could not appear as it does.

The total amount of mass in the universe, similarly, is astonishingly close to the exact amount required for a closed universe. If it had a few times more mass, then gravity would have long since collapsed the universe; a few times less and matter would have dispersed much more widely, likely preventing the formation of galaxies and stars.

Perhaps, when we understand more of the subtle balances between matter and forces—when we have our TOE—we will find that all the universal constants that describe the properties of matter (fermions) and forces (bosons) are preordained. Perhaps only one internally consistent set exists, and because we are here to ponder the cosmos, that is the set we observe.

STUFF:
HOW DO
ATOMS
COMBINE?

In the late 1960s, astronomers detected evidence of mysterious molecular compounds concentrated in wispy clouds of interstellar gas and dust that lie hundreds of light-years beyond the solar system. These substances, distinguished by the way they absorb the light streaming from more distant stars, appear rather similar to carbon-rich soot—a trait that led one researcher to call the effect "auto exhaust along the Milky Way." But the clouds contained something quite unlike anything known on Earth.

Intrigued by the astronomical puzzle, a small group of chemists in Europe and the United States began to study molecular components of soot, hoping to find a match. They vaporized graphite, a form of pure carbon, with an intense laser beam and passed the resulting molecular fragments through a machine that sorted them according to their mass. Over and over again a sharp peak corresponding to an atomic weight of 720—a molecule of exactly sixty carbon atoms or C_{60}—appeared in the spectra, along with many other more and less massive components. What could this unknown collection of carbon atoms be?

Researchers were challenged by the complex jumble of dozens of different carbon-rich molecules in their laboratory soot, which made it difficult to study the specific characteristics of C_{60} in any systematic way. Without a purified sample and a definitive description, most chemists paid little attention to the findings. But in 1990 German scientists learned to produce large quantities of the C_{60} molecule by vaporizing carbon with a strong electric current in a helium atmosphere. They used benzene to dissolve C_{60} in the soot, which formed a beautiful wine-red solution. As the benzene evaporated, beautiful crystals of pure C_{60} appeared.

In a matter of weeks, as news of the C_{60} crystals spread, thousands of scientists joined the research effort. The new and completely unexpected form of carbon was found to possess an elegant soccer-ball-like structure with sturdy cross-bracing bonds arranged like the girders of Buckminster Fuller's geodesic domes. Hundreds of research papers described remarkable physical properties and chemical variants of these fullerenes, or "buckyballs" as the new molecule was quickly dubbed.

Even as scientists marveled at the beauty of the new molecule, a flood of speculative ideas for futuristic applications began to appear. Buckyballs could serve as atomic-scale lubricants, catalysts, chemical filters, and even as an exotic rocket fuel. The carbon cages are tough and resilient, and can team up with other atoms in remarkable ways. Add a bit of potassium and they become superconducting; add fluorine and they act as super-smooth atomic-scale ball bearings. Dangerous radioactive atoms can be placed inside the 60-carbon cages, perhaps providing a new approach to storage or providing new delivery systems for radiotherapy. "Buckytubes," atomic-scale geodesic cylinders of carbon, may offer the first commercial applications for the new family of materials. These incredible fibers, lighter and stronger than any other known filament, could usher in a new era of architecture and design. In all this initial hyperbole, one medical research group even suggested that buckminsterfullerene might inhibit the AIDS virus.

Buckyballs, which garnered the 1996 Nobel prize in chemistry for its codiscoverers, provide the kind of discovery that makes science exciting. No one predicted this discovery, much less its profound impact on chemistry, even after centuries of intense study of carbon in its other two known crystalline forms, graphite and diamond. The textbooks are literally being rewritten.

In a way it's rather disturbing that such an important discovery could be so completely unanticipated. After centuries of research, do we really know so little about atoms?

THE AGE OF
ATOMIC ARCHITECTURE

Archaeologists define each great age of early human civilization by its most sophisticated material: The Stone Age, the Iron Age, and the Bronze Age delineate a slow but steady progression in the creation of implements for commerce and war. Colonial American artisans two hundred fifty years ago relied on no more than a couple of dozen different materials, principally wood, rock, leather, bone, natural fibers, pottery, glass, and a few simple metals.

An explosion of chemical discoveries in the nineteenth century changed our material world forever. The battery, invented by Allesandro Volta in 1800, was instrumental in the separation and identification of many new chemical elements. These findings led to Dimitri Mendeleev's breakthrough discovery of the predictable, periodic properties of atoms and his familiar table of elements in 1869. The periodic table became the chemists' playground, and combining elements in novel ways became their favorite game.

In their exuberant rush to develop new dyes, drugs, fuels, fabrics, fertilizers, glues, paints, preservatives, pesticides, packaging, inks, insulation, cements, cleaners, cosmetics, and countless other products, scientists began to recognize and articulate general patterns in the behavior of the elements and their chemical reactions. Today, after almost two centuries of research, the underlying principles of the science seem to be well established. Every chemistry student learns numerous rules about the sizes and reactivity of elements, the structures and properties of chemical compounds, and the products and rates of chemical reactions. Thanks to these guiding principles, which are now used by thousands of scientists in their efforts to craft compounds atom by atom, a new era of atomic-scale architecture has arrived.

Chemistry is a discipline of elegance and beauty. Its practitioners revel in the versatility of atoms, the esthetics of crystal structures, and the challenges of inventing substances that no human has ever seen before. While quarks and quasars may steal science headlines, it is the new materials developed by chemists that make the big bucks. For many chemists and the corporations for whom they work, the joy of scientific discovery is augmented by the knowledge that materials are indispensable to modern society. We live in an age of extraordinary new materials—substances undreamed of a few decades ago. A succession of revolutions in steels, plastics, semiconductors, and optic fibers has subjected society to a whirlwind of change. New lightweight metals, shatterproof glass, superslick lubricants, and halogen headlights make our automobiles safer and more efficient. Fiberglass insulation, polymer plumbing, silicone caulking, and latex paints make our homes cheaper and easier to maintain. We value these diverse materials for their immediately useful properties. And when a material has properties that are useful, someone will try to find a way to make it better, faster, or cheaper.

Questions about the startling variety of chemical combinations and their properties have implications for our existence that are as profound as any quest for knowledge we can mount. All of matter consists of an organized collection of atoms. Atoms in combination become life—they live and breathe, laugh and love.

Why, then, does chemistry lack glamour? In spite of the paramount importance of materials in our lives, mixing atoms has little of the scientific panache of smashing them. Compared to the countless headlines and best-selling books on particle physics and cosmology, relatively few articles or books for nonspecialists recount the seminal discoveries of chemistry or outline the future promise of materials research. Why is this?

Perhaps because chemistry exists at a familiar scale, unlike elusive subatomic particles or dramatic exploding stars. Chemical reactions such as cooking, washing, gluing, burning, eating, and myriad other activities are utterly commonplace. These phenomena are well understood at the atomic scale, and thus lack the mystery of the unknown universe at its extremes.

In a recent editorial in *Chemical and Engineering News*, the profession's most widely read weekly journal, editor Madeleine Jacobs lamented the public's indifference. Jacobs noted the common view that modern chemistry is "a settled area of science," one which is no longer in an "explosive phase of discovery," like cosmology or molecular biology. Does chemistry no longer address a compelling, fundamental unanswered question?

If chemistry today were little more than routine mixing of chemicals to produce new products, then it would be no surprise that it is perceived by outsiders as boring. It's all too easy to think of chemists as content simply filling in the few remaining chemical blanks, like a club of aged postage stamp collectors. But this perception is wrong for two reasons. Not only does chemical research constantly address a fundamental unanswered question, but its pursuit has also evolved into a creative endeavor of immense beauty and power.

Scientists attach special importance to individual atoms and the minuscule particles of which they are made. This reductionist view holds that nature can be understood through an investigation of its most basic building blocks. In this sense, chemistry has advanced far beyond high-energy physics. While particle physicists still search for a "theory of everything" that systematizes all fermions and bosons, chemists have enjoyed intimate knowledge of their complete set of fundamental particles, the chemical elements, for more than a century. In the periodic table of the elements and the mathematical formalisms of quantum mechanics, they have found an elegant unifying theory. Chemists can predict in great detail how atoms will bond to one another, as well as many of the physical properties of the resulting compounds.

What chemists cannot yet do, however, is anticipate even a small fraction of the possible complex arrangements of atoms, or their emergent properties. Scientists are outrageously far from predicting the behavior of a river, a rock, or a red wine from basic principles. *The* great unanswered question of chemistry and materials science remains: In what ways do atoms combine to form the astounding diversity of objects that enrich the universe? The quest is never-ending, for virtually every chemical experiment adds a tiny piece to this vastly complex puzzle.

Nobel prize–winning chemist Roald Hoffmann argues that chemistry, in this context, transcends the traditional scientific metaphor of exploration and discovery. The number of possible combinations of atoms is for all intents and purposes infinite. Not only do today's chemists discover and duplicate compounds that nature has made, they also imagine and create entirely new molecules, much as a sculptor or architect creates a three-dimensional work of art. Like any creative process, chemistry requires a firm grounding in the rules that govern its medium—in this case the principles of interatomic bonding and chemical reactions. But great chemists also bring daring, imagination, intuition, and passion to their craft. Their finest work displays the same aesthetic refinement and deep beauty of any art form.

What the Theory of Everything Doesn't Tell Us

The theoretical physicists' dream of a theory of everything is not a vision of the end of science. Even a theory that establishes all of the universe's building blocks cannot predict the extraordinary range of objects in the natural world, much less their remarkable properties in combination. Material science, which attempts to bridge the vast gap between descriptions of individual subatomic particles and the behavior of the everyday world, is in essence the study of collective properties of atoms.

The reductionist view of the universe attempts to explain all of the natural world by reducing it to its most fundamental pieces: Animals are made of organs, organs of cells, cells of organelles, organelles of molecules, and so on. This top-down approach has proven invaluable in describing many aspects of the natural world. But complex physical systems, including lava flows, spiral galaxies, and living things, exhibit behaviors that are not obviously programmed into quarks and leptons.

Atoms, like people, may exhibit extraordinary and seemingly unpredictable behavior when gathered together in bunches. Fire, wind, dew, lightning, stars, and life require trillions upon trillions of atoms working in concert. "Fuzzy," "blue," "wet," "crystalline," "hot," "sticky," "superconducting," and "sweet" describe collective properties quite unlike anything associated with individual atoms. Scientists have recently recognized an entirely new class of scientific questions, grouped under the term "complexity," that address such collective behavior.

Consider a snowflake. The internal bonding of one water molecule to another is precisely dictated by quantifiable interactions between the electrons of hydrogen and oxygen atoms. The external form of a snow-flake, on the other hand, is the result of random growth patterns, and is thus an intrinsically unpredictable macroscopic example of that bonding. By the same token, the materials we call life form from relatively simple building blocks—carbon-based molecules with well-defined physical and chemical properties that are predictable from fundamental principles. In life, however, these molecules combine in extraordinary self-replicating assemblages. Scientists may never be able to predict the behavior of living things from fundamental principles alone. A different, more empirical approach is needed to answer chemistry's central question.

How Chemists Answer Questions

Chemical rules are immensely successful at systematizing the properties of known compounds as well as their reactions to form new ones. But because no existing theory allows us to predict all the ways atoms can combine, chemists often mix elements systematically at the lab bench to see what happens. Gradually over the decades, as countless thousands of atomic combinations are tested, old rules are modified and additional organizing principles emerge. And almost any new material discovery holds the possibility of a useful application.

It might appear that a process of discovery as simple as mixing and heating elements would have been exhausted decades ago. After all, the periodic table has only about a hundred different chemical elements, and hundreds of thousands of chemists have worked on the problem. But it turns out that the task is so vast that it will never be completed.

The simplest compounds are made from only two different elements. Combining more than one hundred different elements with each other in a one-to-one ratio of atoms results in more than five thousand possible mixes. And chemists also need to try many other ratios of elements, resulting in hundreds of thousands of mixtures of two different elements. Each of those combinations needs to be studied at many different temperatures, with different cooking times and cooling rates. Changing temperature produces new states of matter as solid melts to liquid and liquid boils to gas. Chemists can also vary the pressure at which the reactions take place, similar to using a pressure cooker. In fact, the whole process is a lot like inventing new recipes, with millions of recipes to try.

But that's only the barest of beginnings. Some pairs of elements, such as hydrogen and carbon, produce an almost infinite variety of compounds. In hydrocarbons, each carbon atom easily bonds to four other atoms; natural gas or methane, for example, features a carbon bound to four hydrogen atoms. Larger molecules result when carbon atoms link to each other to form a backbone: the common fuels ethane and propane have chains of two and three carbons respectively.

The real fun begins with butane, which has four carbons and ten hydrogens. In "normal" butane, the four carbon atoms line up in a row. But the carbon atoms can just as easily adopt a T shape, with a central carbon linked to the other three, as occurs in isobutane. These two forms of butane contain the exact same combination of atoms, but in very different geometries that lead to different physical properties.

Adding more carbon atoms in the chain increases the number of straight, branching, or ringlike structure variants, or isomers. Octane, an important component of gasoline with eight carbon atoms, has eighteen different isomers (and the right mix is essential to the smooth operation of your car, hence the octane rating). Larger hydrocarbons have huge numbers of isomers: Thirty-carbon molecules have billions of isomers, while forty-carbon molecules may occur in sixty trillion isomers.

Many everyday materials have more than two elements in combination. There are about ten million combinations of four different elements, and many trillions of compounds when these elements are combined in different ratios. By one recent estimate, less than one percent of all possible four-component systems has been studied systematically. What's more, many useful materials rely on just a trace of one or more elements for their valuable properties. An impurity of as little as one atom in a billion can drastically alter the strength of glass, the color of gemstones, or the electrical properties of semiconductors. Even with centuries of testing, the number of combinations and conditions left to explore is, for all practical purposes, infinite.

Nonetheless, mixing chemicals is not a completely random pursuit. Systematic trends in the way elements and compounds behave allow chemists to predict accurately what will happen most of the time. For example, the majority of elements are malleable metals with good electrical conductivity and a shiny luster. Mix two metallic elements together and the chances are pretty good that an alloy with similar metallic properties will result. Mix a metal with oxygen, however, and you'll probably get a hard, brittle material like pottery, glass, or rock. Trends of this sort guide research, but nothing can replace the actual mixing of chemicals to see what happens.

Most scientific questions have short answers: "How big?" "How much?" and "How far?" Unknown objects can be identified with one word or phrase. And many natural processes can be modeled with a single equation. In spite of the stereotype of scientists, who are often portrayed as giving long-winded responses to short questions, the language of science facilitates accurate, concise answers.

Some scientists envision a time when many of the deepest questions of science have been answered. Many cosmologists believe that we can know the past, present, and future state of the universe; geophysicists imagine Earth models of exquisite detail; biologists hope eventually to understand the secrets of life itself; and physicists talk of explaining all of existence in a single universal equation.

Yet, even if these dreams come true, thousands of chemical research teams will continue in their quest to understand how atoms combine, as they devise new materials.

ATOMIC ATTRACTION

The diverse properties of the materials that surround our lives result from the strength and arrangement of bonds between atoms. The discovery and systematic description of chemical bonding constitutes one of the great advances of modern science.

When two atoms approach each other, their swirling clouds of electrons come into contact long before the nuclei have a chance to interact. Electrons and their behavior thus represent the key to understanding atoms in combination. Atoms join in chemical bonds primarily because of a curious feature of electrons: The energy of an atom is much lower when its electrons form complete "shells." The first shell of every atom can hold a maximum of two electrons, the second and third shells eight electrons, and the fourth and fifth shells eighteen electrons. Atoms with magic numbers of electrons, two, ten (2 + 8), eighteen (2 + 8 + 8) and so forth, corresponding to filled shells, are unusually stable. Atoms lacking these magic numbers often attempt to add or subtract electrons from their surroundings in an effort to match one of them. Thus, when atoms with eleven electrons (the soft, silvery metal sodium) contact atoms with seventeen electrons (the poisonous yellow-green gas chlorine), a bright explosion occurs as electrons rush from sodium to chlorine and energy is liberated. Ultimately, ten electrons surround each sodium, eighteen electrons fill the three shells around each chlorine, and a bit of table salt (sodium chloride) appears.

Atoms adopt three primary strategies to adjust their number of electrons. Ionic bonds form when one electron or more is transferred from one atom to another, as in the case of sodium bonding to chlorine. Covalent bonds occur when electrons are shared equally among a group of atoms, a situation common in plastics and semiconductors. And metallic bonds arise when atoms relinquish some of their electrons, which are then free to wander about without any home base.

With the notable exceptions of radioactivity and to a lesser extent density, which arise from characteristics of atomic nuclei, most of a material's familiar properties are a consequence of electrons: how they are distributed and shared, the strength of the resulting bonds, the way they interact with light, and so forth. The most immediately obvious property of a material is its state: gas, liquid, or solid. In gases, individual atoms or molecules are so weakly bonded to each other that they fly off to fill any available volume. In liquids, bonds form and reform as the flowing fluid fills whatever shape is available. Bonding between atoms in solids, by contrast, is strong and directed to define a fixed shape and volume.

Everyday solids are further divided into crystals, glasses, and plastics, depending on the regularity of their bonds. Single crystals, including most gemstones and individual grains of sand or salt, feature regularly repeating patterns of atoms. If you could be shrunk to a ten-billionth of your present size and set down on one atom in a crystal, once you got your bearings you could quickly locate any other atom, just like moving around a regular gridwork of city streets.

Glass often looks "crystal clear," but its atomic arrangement is irregular, far from the strict regimentation of crystalline matter. If you found yourself on an atom in a glass, you'd have difficulty predicting the location of any atom more than one or two bonds away.

Plastics, which future archaeologists may identify as the defining material of our present age, represent an intermediate structural state between crystals and glass. They form from long regular chains of molecules called polymers. Moving along any one polymer chain, you would find a regular atomic pattern like beads on a string. But the polymer chains are completely intertwined in a random tangled mass. Hot plastic becomes soft because polymer strands slide by each other; thus softened, it can be molded and cooled into new, useful shapes.

The distribution of strong and weak bonds in crystals, glasses, and plastics leads to many of their most distinctive properties. Consider diamond and graphite, both pure carbon. All the carbon-carbon bonds in crystalline diamond are exceptionally strong and rigid. These bonds are arrayed in a dense framework, like cross-linking beams of a railroad trestle, leading to the hardest known material. But graphite, a layered structure with a mixture of different kinds of carbon-carbon bonds, is only as strong as its weakest links. Carbon-carbon bonds linking atoms within carbon layers of graphite are actually stronger than those in diamond. The other bonds that hold these layers together, however, are extremely weak, making graphite one of the softest known minerals and an effective lubricant. Recently, materials scientists have discovered remarkable fibrous forms of carbon, which feature long chains of strong carbon-carbon bonds. These carbon filaments provide engineers with flexible fibers of unparalleled strength.

Virtually any property of matter—color, luster, electrical conductivity, melting point, magnetic characteristics, and dozens of others—arises from the way atoms combine. Predicting these collective properties can be notoriously difficult. The strength of steel may increase a thousandfold by heat treatments and slight chemical modifications, but it can decrease a corresponding amount if microscopic defects are present. Substitution of one atom in a million by aluminum or phosphorus increases the electrical conductivity of silicon crystals by many orders of magnitude. A few chromium atoms turn colorless aluminum oxide into gem-quality ruby of the deepest red.

No wonder, then, that discovering new materials seems at times as much an art as a science.

Serendipity and Sweat:
How Do Chemists Discover New Materials?

In their quest for novelty, chemists adopt many creative strategies, all of which depend on mixing the right atoms at the right conditions. The simplest chemical reactions are a lot like cooking. Mix a few chemicals, heat them up, and see what happens. Chemists use this "cook-and-look" approach much of the time. Countless useful materials, including Teflon, Velcro, and the sticky stuff on Post-it notes, were discovered quite by chance through such basic benchtop chemistry. Charles Goodyear recounted the accidental discovery of "vulcanized" rubber in 1834, when a sample of gummy natural rubber, "being carelessly brought in contact with a hot stove," turned into a tough black residue. A similar event may have taken place thousands of years earlier, when a Stone Age cook inadvertently produced metallic lead by using chunks of ore to line a fire. All it takes is a bit of luck, and a good eye for something new and useful.

This serendipitous search for new materials is not unlike life's process of natural selection. Nature introduces countless millions of random mutations. While most of these alterations confer no advantage, once in a while a modification represents a real improvement that may win out in the struggle for ascendancy and a new variety of life will be introduced. So, too, are new materials created in vast numbers; the relatively small number that represents real improvement over existing materials may survive as the next generation of products.

In chemistry, as in life's other pursuits, simple methods are often the best. Nothing is simpler than mixing chemicals and heating them on a burner, and most chemistry is still done essentially that way. But atoms often must be coaxed into new configurations by more sophisticated means. The semiconductor industry produces microchips by releasing silicon and other atoms one by one into a vacuum so that they can settle, layer by atomic layer, into the exact desired pattern. Industrial facilities synthesize advanced abrasives by subjecting atoms to crushing pressure in giant viselike machines. Other specialized processing of materials takes advantage of strong electrical and magnetic fields, violent shaking and mixing, or the microgravity of space to produce distinctive collections of atoms.

In the words of the late master chemist Thomas Hoering, "Whatever works, works."

FUTURE WORLD

Fundamental questions notwithstanding, one of the most delightful and rewarding facets of chemistry is the invention of completely new stuff. What wild new materials will society use a century from now? Steel may be replaced by superstrong lightweight carbon filament construction materials for mile-high skyscrapers built with transparent, flexible metal windows that never shatter. Automated cars may float to work on magnetically levitated superconductors. Artificial muscles may help sufferers of degenerative diseases, while artificial nerves repair spinal cord injuries. Dentists may use adhesives that stay rock-hard for years, but can be dissolved away in seconds when necessary. It's almost impossible to comprehend the seemingly infinite variety of materials we have at our disposal, much less predict what exotic new substances await discovery, but here are a few likely prospects that are on the drawing boards right now.

Superconductors

Not too many years ago, newspapers publicized a marvelous idea: a jet train, magnetically levitated by superconductors above its tracks, speeding hundreds of miles per hour on a smooth cushion of air. Commuters would ride from downtown Washington to central New York City in forty-five minutes; cross-country express trains, New York to Los Angeles, would take four hours.

The superconductor hype, based on a strange class of materials that conduct electricity without any resistance or loss of energy, promised a lot more than it has delivered. Cheaper electricity, frictionless motors, electricity storage rings, and levitating bullet trains have not yet transformed our lives, nor are they likely to do so in the next decade. To be sure, superconductors find important specialized applications in medical diagnosis, military sensors, and sophisticated apparatus of physics laboratories, but these technologies rely on materials that must be cooled to near absolute zero with complex refrigeration hardware.

The wave of "high-temperature" superconductor excitement hit the headlines in 1987, when a team of scientists in Texas and Alabama trumpeted their discovery of a brittle black compound that became superconducting at a relatively balmy −170° C—warm enough to rely on cheap refrigerants. Gradually, researchers have pushed that temperature to as high as −110° C, and some optimists still talk about the elusive goal of room-temperature superconductivity. But nagging problems remain. All the new superconductors are as brittle as pottery, not the kind of stuff that's suited for making flexible wire. Furthermore, these materials are extremely unstable, breaking down to a useless black powder when exposed to water or humid air.

For high-temperature superconductors to be useful widely, at least one of three things must happen. Superconductor sales would soar if scientists could discover new, cheap ways to fabricate the materials into efficient, long-lasting products. Alternatively, if costs for electrical energy increase drastically, giant superconducting storage rings that store electrical energy during low-use nighttime hours and distribute energy during peak daytime demand might become economical. And if scientists could discover a room-temperature superconductor that needs no refrigeration, all aspects of electrical supply and use might be transformed. We are still a long way from developing such a material, but every year hundreds of chemists synthesize tens of thousands of new compounds in five-, six-, and even seven-element systems in the search.

But high-temperature superconductors, even if they become commonplace, will probably not change our lives in any profound way. Consumers quickly take advanced ceramics, alloys, and plastics for granted as they become familiar backdrops to living. If superconductors do play a major role in modern technology, they will probably remain hidden from view, quietly allowing us to do whatever we do a little faster or a little cheaper.

Bionics: The Search to
Develop Materials for Life

The age of bionic men and women is on its way. Evolution has resulted in an astonishing variety of lightweight, resilient materials. Muscles, tendons, skin, and hair are, pound for pound, as strong as steel, while teeth and bones are tough and durable under a lifetime of stress. Biochemists are now attempting to copy nature by developing synthetic substitutes for these materials of life.

One key to rebuilding the human body is polymers. Most of nature's materials are modular, composed of a few simple molecular building blocks that repeat over and over again like bricks in a wall. Polymers, which are built from long chains of a simple molecule, are among the most versatile of these materials, forming spiderwebs, plant fibers, tendons, and blood clots. Scientists mimic nature by designing new polymers: nylon, rayon, polyethylene, polyvinyl chloride, and thousands of others.

Remarkable progress is being made on many fronts in attempting to synthesize biological materials. Bioengineers at the Massachusetts Institute of Technology have discovered a polymer that attracts only liver cells—a key step in creating an artificial liver. Researchers at the Buffalo campus of the State University of New York have designed polymer templates for regenerating nerves. And polymers have been used successfully to replace skin, blood vessels, and heart valves. While all of these advances are in the very early stages, and many obstacles remain, medical engineers may someday learn to replace many of the body's structures.

Most polymers are reasonably good electrical insulators, which is why electrical cords and wall switches are covered in protective plastic. Conventional wisdom claimed that polymers couldn't transfer electricity. But a new generation of remarkable polymers may carry electrical currents in our bodies. The discovery of a conducting polymer was made quite by accident about twenty years ago, when a Japanese researcher mixed a thousand times too much of one ingredient during a routine experiment. The new silvery plastic sheet he created had unexpected electrical properties. Today dozens of conducting polymers are known, some approaching household copper wire in their ability to transmit electricity (though none is yet as cheap to manufacture as copper).

Thin, flexible conducting polymers may someday be used to repair damaged nerve fibers. Other electrostrictive polymers, which contract when an electrical current is applied, might provide artificial muscles. A bundle of these polymer fibers could be shaped to look and behave just like real muscles. Not only would this material lead to entirely new treatments for degenerative muscle diseases, but it could also introduce a new era of muscle-enhancing cosmetic surgery.

Superstrong artificial tendons, ligaments, and muscles could be augmented by artificial bones as well. Of all the structural materials in the human body, none is more complex than bone, a composite material that combines the rigidity of pottery with the strength of polymer fibers. Composite materials play many roles in modern life. Automobile safety glass sandwiches a flexible plastic sheet between two brittle plates of glass. Reinforced concrete for road and building construction features flexible steel rods embedded in a durable, rocklike matrix. State-of-the-art carbon fiber composites, which find use in a growing variety of products from jet aircraft to golf clubs, employ finely spun carbon threads in a solid mold of epoxy resin. All of these traditional composites rely on relatively large components that can be held in your hand—metal rods and glass fibers, for example. But, as in so much of modern chemistry, the focus of such advances has shifted to a much smaller scale.

The next generation of composite materials, inspired by nature's most resilient materials, will look very different. Living organisms form teeth, bone, and shell by interlocking tough polymers with a rigid ceramic framework at the microscopic scale. Pottery and polymers are both strong in their own way, but each has its limitations. When pushed too far, pottery shatters and polymer fibers buckle. Neither of these materials alone is adequate for the stresses life places on the skeleton, but intimately intergrown, they combine into bone.

Now materials scientists are attempting to imitate these remarkable lightweight, strong composites from raw materials as cheap as coal, air, and water. Artificial bone, made by merging space age ceramics and fibers like a microscopic reinforced concrete, could be almost unbreakable. If manufactured on a large scale, such a material could revolutionize industry and commerce, and find a thousand new uses.

Liquid Crystals

New materials sneak up on us, insinuating themselves into our lives in such ordinary ways that we often don't even notice their arrival. One such material is liquid crystals. Most everyday materials are solid, liquid, or gas, but liquid crystals have very long, slender molecules that can display a behavior intermediate between solid and liquid. The molecules can adopt the normal liquid's structure with completely random orientations, or all the molecules can line up in the same orientation, like individual straws in a box. In such a liquid crystal, the molecules can flow past each other, as in a liquid, but they are precisely oriented as in a crystal.

The utility of liquid crystals arises from their unusual optical properties. Many of these liquids are clear when their molecules are randomly oriented, but opaque or colored when the molecules line up. Furthermore, when such molecules are sensitive to an electric field, they may be made to line up under the influence of a small electric current. The remarkable liquid crystal displays (LCD) in your watch, pocket calculator, and laptop computer work in this way.

Someday the walls of your office or home may be covered in LCD sheets. With the push of a button, the entire wall would display the latest weather map, stock quotations, or your favorite sports team in action. The display could also be hooked up for video conferencing, a face-to-face chat with a distant friend, or perhaps a relaxing panorama of a tranquil lake. Someday users will take this technology for granted, and barely give a passing thought to the liquid crystals that make it possible.

Are These New Materials Safe?

DDT, PCBs, CFCs, Alar, dioxin, asbestos, benzene, radon—the list of chemical hazards grows and grows. Every month researchers warn of new carcinogens, pollutants, and other chemical dangers. No image of modern technology is more damning than a chemical plant spewing noxious fumes into the environment. "Contains absolutely no chemicals!" has become a persuasive if ludicrous advertising slogan for "all natural" products.

The government has taken on the responsibility of regulating the tens of thousands of new chemicals synthesized every year. Of greatest concern are new pesticides, food additives, cosmetics, explosives, radioactive materials, and drugs, which undergo years of rigorous testing and approval processes by a variety of federal agencies. But what about the two thousand or so industrial chemicals introduced to the marketplace annually?

In an effort to protect workers and consumers from dangerous new chemicals, Congress instituted the Toxic Substances Control Act in 1976. The act, administered by the Environmental Protection Agency, requires chemical manufacturers and importers to provide ninety-day advanced notice and obtain EPA approval before marketing any new material. The great majority of these new chemicals are safe—little more than variants of everyday plastics, dyes, adhesives, or other common substances. Each year a few dozen new chemicals require special regulations, and a few potentially dangerous substances are withdrawn from consideration.

Inevitably, some seemingly benign new chemicals will prove harmful, just as they have in the past. It took researchers decades to realize that CFCs, cheap and efficient chemical refrigerants that are completely harmless to humans when breathed, are gradually destroying the Earth's protective ozone layer some thirty miles up in the atmosphere. Similarly, petroleum products with PCBs, pesticides with DDT, and construction materials with asbestos were marketed for many years before their health dangers were fully realized. "We're regulating in the dark," says one EPA official. "We don't know if the program is working."

New chemicals provide untold benefits to society, but they also pose risks. No one can say for sure that one of this year's crop of approved substances might not have consequences similar to those of CFCs or PCBs decades from now. As long as society demands new and improved products, it will have to learn to monitor and minimize the risks.

VIRTUAL ATOMS

For centuries, advances in chemistry have been accomplished by mixing chemicals and seeing what happens. Now a growing segment of the chemical community relies on theoretical models, supercomputers, and virtual atoms to calculate the structure and properties of matter. Computers are marvelous at answering questions; the only drawback is that each question must be asked in the form of an equation. The sophisticated art of computer modeling relies on developing a theory of matter and then devising equations that mimic the real world.

The energy released when two or more atoms bond together can be described by a precise mathematical formula. Computer modeling of chemicals relies on equations that calculate this energy—as atoms are brought together, the most stable chemicals produce the largest energy release. Unfortunately, no one has found a suitable energy equation for all materials, and even approximate solutions can take many hours of computer time. Theorists struggle to devise computer programs that are realistic both in terms of the way they describe atomic interactions and in the amount of computer time they consume. As a result, even the most sophisticated efforts provide only an approximation to the physical reality of materials.

Using computers to do chemistry depends on an understanding of the forces that bind atoms together. As a collection of atoms link up, their electrons interact in ways that change the total energy of the system. Any solution for which atoms settle into a new pattern of low energy represents a plausible new material. Furthermore, many properties of that hypothetical material, including electrical behavior, strength, hardness, transparency, and much more, can be calculated, based on known relationships between structure and properties.

It may at first seem odd to use a computer when chemicals can be mixed in the lab, but computer-modeled materials have vast potential to simplify the chemist's task. Computers are safe and clean; they don't get sticky, explode, or release toxic fumes. Computers, and the programs they run, are getting much faster, and might eventually allow one researcher to examine thousands of imagined chemical compounds every day. Furthermore, computers are not constrained by a laboratory's physical limitations; they can model matter at extremes of temperature and pressure far beyond any present experimental setup, including conditions corresponding to the deep interiors of planets or stars.

Most important, theoretical models help material scientists understand *why* chemicals behave the way they do. Computers will never replace experimentation, but we have reached the point where scientists are able to make specific predictions about simple systems based on their computer analyses. Researchers at Corning, IBM, and several major pharmaceutical companies, for example, now routinely use computer-aided materials design, much as aircraft manufacturers test their plane designs on computer before building a full-scale version. Of course, once a promising substance is predicted, an actual experiment with real atoms must be performed.

The promise of computational chemistry is exemplified by the search for superhard materials. Physicist Marvin Cohen, a professor at the University of California's Berkeley campus, took a scientific gamble in the 1960s when he began to use computers to predict the behavior of new materials. At that time few people took his work seriously. "It was a joke. We were second-class citizens," Cohen recalls of the initial unenthusiastic reception for his studies. But sometimes gambles pay off.

Cohen models solids with a computer in a lifelong effort to understand how atoms combine to make reality. In decades of groundbreaking research, he has developed ever more sophisticated computer programs to model atoms and their electrons at the quantum level. The simple models of the 1960s, now viewed as primitive as dinosaurs, could incorporate a handful of atoms at a temperature of absolute zero, where vibrations are minimal. As computers and their programs have increased in complexity and speed, more complex and more realistic problems have become viable.

In one of his most publicized efforts, Cohen and his student Amy Liu in 1989 predicted the existence of a new material that may rival diamond in hardness, a then-unknown compound of carbon and nitrogen, C_3N_4. With Cohen's model as a guide, in 1993 researchers at Harvard University successfully synthesized a thin film of this material, and patents on the potentially valuable substance were filed shortly thereafter. Many other researchers, following Cohen's lead, continue to explore this promising approach to superhard materials. Cohen and others predict that materials ten percent harder than diamond may be synthesized at high pressure within the next year or two. Whatever the outcome, computer chemists will help us to know a little more about the materials that form our universe.

Computational chemists envision a day in the distant future when not just single compounds, but complex chemical reactions are studied first at the keyboard of a computer terminal. A century from now, we may look back on benchtop, cook-and-look chemistry as the dark ages. Someday theorists may achieve their ultimate dream: Enter nothing more into the computer than the mix of elements to be combined, and out pops a complete description of the material in question.

THE QUEST FOR ENERGY: WILL WE RUN OUT?

Until such time as there are new technological breakthroughs, . . .
industrial society has only three primary clusters of alternatives on
which to rely for its new power needs: oil, gas, and coal; nuclear
power; and conservation.

—Daniel Yergin, *The Prize* (1991)

Few people care about the source of our energy supply except
when it is disrupted.

—Ged R. Davis, *Scientific American* (1990)

If all proceeds as planned, the fabled year 2001 will witness a break-through experiment of awesome implications. A fearsome, bristling array of 192 powerful lasers, each capable of blasting a hole through a block of stone, will focus its energy on a dense BB-sized pellet of hydrogen fuel. Tense teams of scientists and engineers will monitor banks of electrical capacitors, timers, and recording devices as the countdown proceeds. The laser array will fire in precise unison, devastating the tiny fuel pellet with focused radiant energy equivalent to a stick of dynamite. Though the brilliant pulse will last only a few billionths of a second, the tiny sphere will implode violently. Hydrogen fuel will be compressed to densities approaching one hundred times that of lead, as temperatures soar to millions of degrees. At such unimaginable extremes, comparable to conditions deep inside the sun, hydrogen transforms to helium in a thermonuclear reaction—a miniature hydrogen bomb—that releases far more energy than was consumed by the power-greedy lasers.

Such is the dream of researchers at California's Lawrence Livermore National Laboratory, who hope to coax and cajole a half billion dollars or more from American taxpayers to build the National Ignition Facility. The venture is not without risks, but the ultimate payoff is vast, for these scientists seek to harness the virtually unlimited fusion energy that powers the sun.

THE LAWS OF ENERGY
IN OUR EVERYDAY LIVES

The business of life in a physical universe is conducted with matter and energy. Several fundamental questions about matter remain unanswered. What is dark matter? How much matter does the universe hold? What are matter's basic building blocks, and can they be systematized in a theory of everything? Basic principles related to the behavior of energy, on the other hand, are among the best understood and most thoroughly documented aspects of our physical world, and few scientists expect to stumble across any profound new insights. Multifaceted energy research now lies primarily on the pragmatic side of the blurred boundary between pure and applied science.

It was not always so. One hundred fifty years ago, no scientific questions loomed larger than those connected to the mystery of energy. What is it? What kinds are there? How does it behave? Those were fascinating, fundamental questions, and researchers devoted their lives to answering them. Many of their results are summarized in two great laws of thermodynamics—the laws that guide all modern energy research.

The First Law: Conservation of Energy

Energy, though now recognized as a pervasive theme in all scientific fields, is a subtle concept that was not well understood until the mid-nineteenth century. Energy proved difficult to define because it occurs in a variety of forms and is manifest in many different kinds of phenomena. For centuries, the principal kinds of energy that affect our lives—potential, kinetic, heat, and light—were investigated as independent phenomena. (The discovery of nuclear energy, which plays no obvious role in everyday experience, occurred decades after the laws of thermodynamics were deduced.)

"What's out there?" is the logical first question to ask in any systematic study of nature. Naturalists identify and describe related objects or phenomena before they can be understood in any integrated way. If science always progressed in an unswerving march toward truth, energy would have been investigated in this straightforward manner. Like intrepid botanists voyaging to exotic South Seas islands to catalogue plants, clever physicists would have searched high and low for all the different varieties of energy. Unfortunately, the several kinds of energy are not so obviously related to each other as an orchid and a fern.

Sources of stored or potential energy, though not then referred to as such, were quite familiar to observers two centuries ago. The gravitational potential energy of suspended weights and dammed-up water, the magnetic potential energy of a north-swinging compass needle, the chemical potential energy of food and fuel, and the elastic potential energy of a taut bowstring or metal spring were sources of energy widely used in industry and commerce. Two centuries ago scientists also observed that moving objects such as a brisk wind or flying cannonball possess the ability to do work. These objects carry kinetic energy, the energy of motion.

The recognition of heat as an additional source of energy represents a crucial breakthrough in the history of energy research. Key observations were made by exiled American physicist Benjamin Thompson, who fled the United States in 1776 after spying for the British during the Revolution. The resourceful Thompson found employment as master of the cannon works in Bavaria, where, in 1798, he began investigations on friction and heat. According to the theory prevailing at the time, heat was perceived as a form of fluid matter that could be liberated from objects when they were burned, abraded, or otherwise broken apart. Thompson noticed a curious phenomenon as his brass cannon were being bored: Dull tools produce much more heat than sharp ones, even though less metal is removed. In fact, the slower the boring, the more heat produced, so Thompson reasoned that heat could not be a fluid that saturated the brass. Instead, he found that heat arose from the mechanical action of friction.

By the 1850s, several scientists had independently recognized that the familiar phenomena of potential, kinetic, and heat energy readily transform, one into another. Once these diverse varieties of energy were recognized, and their interchangeability confirmed, it did not take researchers long to formulate the first overarching energy principle, called the first law of thermodynamics: The energy in a closed system may change forms many times, but the total amount of energy remains constant.

The first law comes into play in countless ways. Automobiles convert the chemical energy of gasoline into kinetic energy. Hydroelectric power plants transform the gravitational potential energy of dammed-up water into electricity. Photosynthetic pathways in plants change the sun's radiant energy into the stored chemical energy of food. Our bodies convert chemical energy into kinetic energy of movement and the electrical energy of thought. All around us, energy constantly shifts from one form to another.

The Second Law: Nature's Direction

The first law of thermodynamics places no restrictions on energy transformations, but we know from practical experience that energy shifts from one form to another in predictable ways. On a pleasant spring day a cup of hot chocolate spontaneously cools, while ice cubes melt. A taut rubber band snaps, but never spontaneously becomes taut. Objects roll downhill, converting gravitational potential energy to kinetic energy— never the other way around.

Nature's familiar directionality is formalized in the second law of thermodynamics. Simply stated, the second law says that energy always flows spontaneously from a hotter to a cooler body; from more concentrated to less concentrated forms of energy.

Consequences of the second law also pervade our lives. Concentrated forms of energy, such as fossil fuels and enriched uranium, are valuable because they represent the most usable forms of potential energy. A lump of coal will burn, combining with oxygen to produce carbon dioxide, water vapor, and lots of useful heat energy. But the process won't reverse: Heat will not spontaneously induce carbon dioxide and water to form a lump of coal and molecules of oxygen. Consequently, we must pay for fuel to heat our homes in the winter and cool them in the summer.

The second law defines the direction of events in time. Water flows downhill. Rooms get dusty. Supplies of fossil fuels diminish. We grow older. We cannot escape the sobering consequences of the second law.

WHERE DO WE
GET OUR ENERGY?

Nature has surrounded us with prodigious amounts of energy. The sun bathes the Earth in radiant energy equal to 15,000 times the world's total consumption, and heat radiating out from the Earth's core could satisfy all our needs for millions of years. Even nonrenewable supplies of carbon-based and nuclear fuels are adequate for many hundreds of years at present consumption rates. We will not run out of *available* energy anytime soon. The principal concern for society is that at present we have neither the technological means nor the infrastructure to exploit more than the tiniest fraction of this abundance.

The Arab oil embargo of 1973 brought great industrial nations face-to-face with the laws of thermodynamics. In the years leading up to the Yom Kippur Arab-Israeli war of October 1973, petroleum production had barely kept up with demand. The unusually cold winter of 1969–1970 resulted in numerous local shortages of oil and natural gas. This precarious situation reached the crisis stage as Arab nations cut off oil supplies and the price of crude oil shot up 600 percent.

Increased energy production and new efficiency measures have temporarily eased the petroleum shortfall. For two decades, the oil has flowed freely. Once again we find ourselves in a situation where fossil fuels—coal, petroleum, and natural gas—are so abundant and easy to mine that little economic incentive exists for individuals and corporations to switch to alternative sources. But make no mistake, we are still vulnerable.

The first law of thermodynamics points to a powerful strategy in our pursuit of untapped energy sources. Given the vast amount of available energy, we need only discover ways to convert these abundant resources into other more useful forms of energy.

Nonrenewable Reserves and Alternative Energy Sources

Whatever the changing economic and political climate, the scientific and technological questions relating to energy are straightforward: What sources of energy are available to us? And how can we use those sources in the most efficient ways?

Geologists seize every means at their disposal to track down hidden mineral wealth. They discover coal seams and petroleum concentrations with tried-and-true field prospecting, supplemented by seismological profiling, satellite reconnaissance, and exploratory drilling. Ground and air surveys pinpoint uranium deposits by their telltale radioactive signatures. Exact figures on the extent and quality of known resources are often jealously guarded by oil and mining companies, and major repositories may lie undetected in remote wilderness areas or beneath the ice of Antarctica. Even so, experts estimate that inventories of fossil and nuclear fuels could last a thousand years or more at present consumption rates (see Table 1). Although petroleum is a dwindling resource with only a few decades of reserves left, the known coal supplies, which can be converted readily to liquid fuels, are sufficient for several centuries. Moreover, coal reserves are supplemented by vast deposits of oil shale and tar sand. As the price of crude oil rises, these less convenient alternative sources will gain in importance.

In spite of this abundance, two serious concerns accompany the unrestricted use of fossil fuels. The most immediate and well-publicized problem is atmospheric pollution, with its resultant smog, acid rain, and buildup in concentrations of the greenhouse gas, carbon dioxide. Emission controls can substantially reduce smog and acid-forming compounds, but carbon dioxide is an inevitable byproduct of all burning. Environmental considerations notwithstanding, many chemists stress conservation of fossil fuels for a second, very different reason. Carbon-based fuels are becoming an ever more precious chemical resource for manufacturing plastics, drugs, building materials, and a host of other products. As scientists devise more uses for these chemicals, carbon-carbon bonds are simply becoming too valuable to burn.

Nuclear power, though of secondary importance to fossil fuels, ac-

TABLE 1

Nonrenewable Energy Resources

Nonrenewable Sources	Estimated Reserves (in quads)*
Petroleum	12,000
Coal	140,000
Natural Gas	20,000
Oil Shale	20,000
Tar Sand	5,000
Uranium—Conventional	2,000
—Breeder Reactor	120,000

*A quad equals one quadrillion BTUs; annual world energy consumption is now approximately 500 quads. Estimates from "Energy Sources: A Realistic Outlook" by Chauncey Starr, Milton F. Searl, and Sy Alpert. *Science* 256, 1992, pp. 981–987.

counts for about a sixth of the world's electricity. Known reserves of enriched uranium fuel for conventional reactors are adequate for perhaps another century at present consumption rates, but a new generation of efficient, fast-breeder reactors could increase the energy produced from uranium fuels by a factor of 100, ensuring several millennia of power. Such a dramatic boost in output is possible because conventional reactors tap less than one percent of the uranium fuel's stored energy. Much more efficient fast-breeder reactors actually create more radioactive fuel than they consume by initiating nuclear reactions that convert some uranium into highly energetic plutonium. Environmental and security issues associated with the large quantities of nuclear waste and weapons material produced by breeder reactors, however, argue against their widespread adoption by the power industry.

Fortunately, nature has provided us with numerous energy alternatives to fossil and nuclear fuels (see Table 2). Forefront research focuses on devising novel ways to convert phenomena as familiar as wind, rain, and sunlight into electricity or heat. Two of these emerging technologies, photovoltaics and nuclear fusion, convey a picture of the formidable challenges and long-term promise of energy research.

TABLE 2

Alternative Energy Resources

Source	Comments
Geothermal	Energy is extracted from hot water or steam produced underground in volcanic areas such as Hawaii and northern California.
Ocean Sources	
Thermal gradients	The contrast between cold deep water and warm shallow water is exploited in a pilot program at Hawaii's National Energy Laboratory.
Waves and tides	Kinetic energy of moving water is tapped in some coastal areas.
Wind Power	Wind turbines supply one percent of California's electricity. Wind farms in the windswept Midwest could provide half of North America's electricity.
Hydroelectric	Rivers and waterfalls provide a sixth of North America's electricity. Environmental concerns will prevent significant expansion.
Sources Based on Converting Sunlight	
Biomass	Plants can be partially converted to alcohol fuels, in addition to wood and other plant-based combustible fuels.

Solar heating	In colder regions, passive solar heating supplements conventional heating for private and commercial buildings.
Solar-thermal	Direct conversion of sunlight to heat energy by this method relies on curved arrays of mirrors to focus sunlight for steam generation.
Solar-chemical	Chemists are developing systems that use solar radiation to break down water into oxygen and clean-burning hydrogen.
Photovoltaics	Semiconductor devices that convert light directly into electricity are used both in large-scale power plants and in portable devices.
Nuclear Fusion	Thermonuclear conversion of hydrogen into helium plus heat energy, the energy source of the sun, could provide a virtually unlimited source of energy on Earth.

What Are Photovoltaics?

Photovoltaics are remarkable materials that almost magically convert daylight directly into electricity. They hold the promise of achieving energy self-sufficiency with a technology that is environmentally benign and widely applicable. Most solar-energy-based technologies, including hydroelectric power, wind turbines, and coal-powered generators, require intermediate mechanical steps before the sun's radiance becomes useful. Earth's cycles transform the sun's energy into gravitational, chemical, or kinetic energy, all of which are harnessed to drive a turbine that generates electricity. Photovoltaic materials, by contrast, transform light into electricity in a single step, without any complex mechanisms.

Photovoltaic devices exploit a thin layer of semiconducting material, such as silicon, that has been bonded to a metal foil. A hallmark of all semiconductors is that some electrons are loosely bound to their home atoms and are ready to shake free at the slightest provocation. When photons of sunlight strike the semiconductor, a few electrons are bumped out; they flow into the metal foil to create an electric current. Unlike other solar energy technologies that require large power-producing facilities, photovoltaics rely on an atomic-scale phenomenon. Solar cells can be miniaturized to any size, small enough to operate a light meter or pocket calculator.

Photovoltaics already find many specialized uses. Thin solar panels covered with shiny black silicon wafers form the familiar "wings" on many satellites, and they are becoming an ever more common architectural feature as well. Pilot plants in the southern California desert employ acres of photovoltaic panels to generate significant amounts of electricity, while a proposed full-scale plant in the Nevada desert will supply enough electricity for 100,000 consumers. Even so, photovoltaics are still in their infancy, and significant advances in efficiency and reductions in cost are needed to realize their full potential.

One of the most intensive, ongoing photovoltaic research efforts takes place at the National Renewable Energy Laboratory in Golden, Colorado. This state-of-the-art facility, sponsored by the Department of Energy, employs more than a thousand energy scientists, engineers, and support staff with an annual budget of a quarter of a billion dollars. Their photovoltaic research, which integrates the efforts of about one hundred solid-state physicists, computational chemists, and materials engineers, concentrates on two key problems: creating new semiconducting materials and devising advanced composite designs.

The prime objective of the laboratory's team is to develop a high-efficiency photovoltaic material that can be manufactured at low cost. A key element of this effort is to synthesize chip after chip of semiconductor material to see which have the best photovoltaic properties. For each chip, researchers must measure many properties: How much of the sun's broad light spectrum is absorbed? How much of that energy is converted into moving electrons? How thin and flexible can the thin layer be? How fast does it degrade and lose its light-gathering ability? At the same time, theorists devise computer models of semiconductors at the electronic level, analyzing why experimental materials behave the way they do, and predicting new combinations of atoms that might work better. Finally, materials engineers take the results of this basic research and devise the rapid and reliable manufacturing procedures necessary to produce acres of photovoltaic panels.

Silicon has long been the semiconductor material of first resort for electronics applications. The raw material is cheap and fabrication technologies have been honed to a fine art. Nevertheless, silicon has several severe limitations. Crystalline silicon, though very efficient in converting close to twenty percent of sunlight into electricity, requires painstaking, labor-intensive fabrication techniques. Silicon atoms must be vaporized in a high-vacuum chamber and deposited, atom by atom, onto an ultraclean surface. Trace impurities, or dopants, perhaps only one atom in a billion of aluminum or phosphorus, precisely tune the light-gathering properties of the semiconducting silicon. With such exacting requirements, crystalline silicon photovoltaic panels are prohibitively expensive for commercial applications.

Rapidly deposited thin films of amorphous (noncrystalline) silicon are much cheaper to produce, but their solar-collecting efficiency is less than ten percent. Amorphous silicon, furthermore, gradually degrades when exposed to strong light for extended periods—not a desirable trait for sun-drenched materials.

Intensive research efforts focus on dozens of photovoltaic alternatives to silicon, as well as variations in dopants and their concentrations. Cadmium telluride, a black semiconductor similar in appearance to silicon, is a stable, easy-to-fabricate material that absorbs a broad band of visible sunlight. Efficiencies exceeding fifteen percent, coupled with long lifetimes and low manufacturing costs, could ultimately rival silicon as the photovoltaic of choice for large-scale power generation. Semiconducting gallium arsenide, by contrast, is as expensive as crystalline silicon to manufacture, but it boasts the high efficiency and stability under extreme conditions that are ideal for the growing market in satellite power applications.

Some of the most innovative efforts in photovoltaic development rely on stacking conventional semiconductors, thus combining the best features of two or more materials. The first layer absorbs the lowest energy infrared photons of sunlight, while subsequent layers successively absorb visible and ultraviolet energy. Scientists at Boeing have produced a two-layer device featuring gallium arsenide and gallium antimonide semiconductors with 38 percent efficiency, while efficiencies exceeding an astonishing 40 percent might be achieved with a triple-layer composite.

Ultimately, the twin factors of efficiency and cost will dictate how pervasive photovoltaics become. Typical commercial designs of the 1990s, featuring a single semiconducting layer of silicon, convert less than 10 percent of the sun's energy into electricity at an initial cost of a few dollars per watt of power—that's a few hundred dollars to light a typical lightbulb. While most consumers are unwilling to spend that much for home electricity, costs will drop as new materials and processing techniques are introduced.

If mass produced at low enough cost, especially if coupled with batteries or other electrical storage devices, a roof or external walls covered with solar cells could easily supply all the electricity for an average household. High-efficiency solar cells could usher in an era of reliable sun-powered automobiles, ships, and perhaps even airplanes. And someday, through the forces of economics and environmental good sense, the vast uninhabitable deserts of the world could be transformed into energy farms adequate to supply all human needs.

Can Nuclear Fusion Be Commercially Viable?

The most fervent dream of many energy researchers is not simply to harness the sun's radiance, but to duplicate it on Earth. The sun's prodigious energy output arises from the conversion of hydrogen to helium in nuclear fusion reactions, during which roughly half a percent of the hydrogen's mass transforms directly into heat and radiant energy. So efficient is this process that a trifling 700 tons of hydrogen consumed each second accounts for all the sun's radiant output into space. And on Earth, the requisite hydrogen fuel is as plentiful, and almost as cheap, as water.

On November 1, 1952, the United States mimicked the sun, unleashing an uncontrolled fusion energy reaction in the first hydrogen bomb explosion. Several other nations followed suit, in what turned out to be a disturbingly straightforward weapons technology. Controlling nuclear fusion in a safe reactor is vastly more difficult. After half a century and ten billion dollars of effort, researchers are still many decades away from a commercially viable nuclear fusion reactor.

Controlled fusion reactions will require heroic scientific and technological advances. Hydrogen cannot spontaneously fuse into helium; a roomful of hydrogen could exist for the lifetime of the universe without a single fusion event ever taking place. To form helium, two hydrogen nuclei with positively charged protons must be brought so close together (about one ten-trillionth of an inch) that the strong nuclear force overcomes the otherwise dominant repulsive electrical force. Hydrogen atoms (or, better yet, their heavy counterparts deuterium and tritium) must be confined in a dense form and heated to temperatures of millions of degrees.

Two contrasting approaches offer promise and are subjects of intense efforts. The most visible and widely publicized path to fusion energy focuses on confining a hot, dense plasma in a strong magnetic field. Plasma is a bizarre state of matter in which electrons have been stripped from their hot host atoms to form a glowing gas of electrically charged particles. Though less familiar than the solid, liquid, and gaseous states that surround us, plasmas form the bulk of all stars and represent by far the most abundant visible matter in the universe. A detailed understanding of the plasma state is thus fundamental to modeling the life and death of the sun and other stars.

One rewarding approach to studying plasma physics has been to corral the ionized gas in a magnetic "bottle." Magnetic fields mold plasma into a dense mass, which can be heated by an electric current, by radio waves (something like a microwave oven), or by collisions with an energetic beam of hot deuterium and tritium nuclei. If the plasma's temperature is sufficiently high, then fusion will take place.

Most magnetic confinement research employs massive ring-shaped devices called tokamaks. The Tokamak Fusion Test Reactor at Princeton, New Jersey, the largest North American facility with an impressive doughnut-diameter of more than sixteen feet, has set many magnetic confinement records on the arduous path to sustained fusion. In 1994, in a landmark experiment, the Princeton facility achieved a commercially viable combination of hydrogen plasma density and high temperature, albeit for only a half second. Other tokamak facilities in England and Japan are now poised to pass the critical break-even barrier, at which point the amount of fusion energy produced exceeds the prodigious energy required to confine and heat the plasma.

An alternative (and until recently top secret) fusion strategy involves blasting a dense spherical pellet of hydrogen-rich material with the energy from an array of high-intensity lasers. This big science process, called inertial confinement fusion, in essence creates a miniature thermonuclear device. In a hydrogen bomb, an atom (nuclear fission) bomb is used to detonate a hydrogen-rich solid such as lithium hydride. The atom bomb heats and compresses the hydrogen fuel to the point of fusion. Inertial confinement fusion achieves exactly the same effect by focusing intense energy beams onto a tiny hydrogen fuel pellet that is surrounded by an explosive surface coating. As the irradiated surface of the pellet explodes outward, the hydrogen fuel implodes and ignites.

The major North American effort in inertial confinement fusion has taken place during the last three decades at the Lawrence Livermore National Laboratory in Livermore, California. The ambitious Nova project, only recently declassified by the government, employed a bank of ten powerful lasers to implode the hydrogen pellet. Nova experiments focused primarily on the vexing problem of compressing pellets uniformly during the violent laser firing; the slightest asymmetry and hydrogen fuel just squirts out sideways. Rather than fire lasers directly at the pellet, the Nova team has devised a clever "indirect drive," in which lasers are focused on a gold capsule surrounding the pellet. As gold absorbs the brilliant laser beams, the metal emits an intense cascade of X rays that floods the capsule, uniformly bombarding the pellet and triggering the desired implosion.

Scientists at Livermore and their collaborators at Los Alamos National Laboratory are eager to scale up Nova with a new fusion experiment, the epic half-billion-dollar National Ignition Facility. Armed with 192 high-powered lasers, each one about as powerful as any now in existence, the facility will focus its immense energy on deuterium-tritium pellets about a tenth of an inch across. If all goes as planned, and Congress approves the hefty price tag, construction could begin in time for test firings in 2001, about the same time that a rival French machine of similar design is supposed to become operational.

In spite of the long-term promise of fusion power, the future of this research in the United States remains uncertain. Ambitious efforts to explore both magnetic and inertial confinement fusion options will require a dramatic increase in funding—by some estimates a tripling of current levels by the year 2000. And even if proposed experiments work flawlessly, researchers are still many decades away from a commercial fusion reactor. A budget-conscious administration may see little immediate payoff in this research. (According to a familiar Capitol Hill joke, fusion energy is the energy source of the future, "and it always will be.") It's hard to imagine how the Department of Energy budget, projected to be cut almost in half in the next five years, can support an expanded fusion program.

THE SEARCH FOR
OTHER OPTIONS

Untapped energy sources provide an obvious path to energy security, but the laws of thermodynamics point to a second effective strategy for increasing available energy. As any thrifty investor knows, you save what you don't spend, and huge energy savings are possible through conservation. Much of the success of these measures will depend on personal choices regarding lifestyles and consumption. Energy savings from recycling of aluminum cans and plastic containers, for example, add up one consumer at a time. Science can also help in conservation efforts by inventing more efficient methods for storing and using energy. While creative research proceeds on many fronts, the modern evolution of the seemingly mundane technology of lightbulbs epitomizes the astonishing originality of efforts to improve efficiency.

Lightbulbs convert electrical energy into visible light. The simple brute-force method we use, favored for thousands of years, is fire: Heat an object until it glows. So it was with the first commercial lightbulbs of the 1870s, when Thomas Edison's creative New Jersey–based research team used an electrical current to heat a filament. So simple and convenient is this design that it persists to this day, with more than a billion and a half incandescent bulb fixtures in the United States alone.

For atoms to emit visible light, their electrons must first absorb energy and jump into excited states. Electrons typically drop back down to their normal state in a fraction of a second, and the absorbed energy radiates away in the form of photons. A bulb's efficiency is defined by the percentage of photons produced in the narrow range of visible light. Modern incandescent bulbs are inefficient because an electric current heats a thin, resistant tungsten wire so hot that energy is emitted primarily as invisible infrared photons (what we feel as heat radiation). The bulb's initial low manufacturing cost is offset by its profligate use of energy and its relatively short 1000-hour lifetime. The trick to improving bulb efficiency is to find an arrangement of atoms whose electrons readily absorb energy and re-emit that energy almost entirely as visible photons.

Fluorescent bulbs, first introduced in the 1930s, pass electric current through an ionized gas of argon and mercury rather than a metal filament. The gas mixture emits most of its photons as invisible ultraviolet radiation, and those ultraviolet photons are in turn absorbed by a phosphor that glows brilliantly on the inner surface of the glass bulb. Fluorescent bulbs remain cool, lasting up to ten times longer than the average incandescent bulb, while using only a sixth the energy to produce the same amount of light. A principal drawback of fluorescent bulbs is their dependence on a starter and ballast, which are required to ionize the gas. These bulky components are not easily adapted to fit the popular screw-in fixtures of incandescent bulbs.

A new generation of remarkable lighting devices promises to be even more efficient and longer lasting than the fluorescent bulb. Recent lightbulb research has focused on alternative methods for exciting electrons. One intriguing device converts electricity into radio waves, like a miniature radio station. These waves excite electrons in the bulb's vapor of mercury atoms, which emit photons of ultraviolet radiation. A phosphor coating glows white, just like in a fluorescent light. Another high-tech bulb uses microwaves to excite electrons in a mixture of argon and sulfur atoms. These new bulbs won't be cheap; manufacturers suggest a price of 15 dollars apiece for the radio bulb. But radio and microwave lamps are projected to last for more than a decade, compared to six months for ordinary bulbs. Furthermore, each bulb could save as much as twenty cents per week in operating costs. All consumers have to do is think of lightbulbs as household appliances instead of disposable items.

WHAT DOES THE FUTURE HOLD?

For most of human history, fire and the sun provided almost all the energy required for heating and lighting. Open flames cooked food, heated dwellings, and illuminated the night. In that fire-dependent society, energy was the daily concern of every individual. Acquiring adequate supplies of firewood and other fuels posed a never-ending, time-consuming chore, while starting, maintaining, and extinguishing fires were skills learned in early childhood.

The electrification of society began in the late nineteenth century and soon imposed a revolution in individual energy use so sweeping and profound that it may not again be equalled for centuries. Electricity, which is now generated at distant power plants, at once isolated energy consumers from the daily rituals of the flame, while for the first time in history it physically linked every household in a community. The independent, individual energy gatherer had been replaced by energy corporations, and the fabric of society was forever changed.

The coming century may see dramatic innovations in the ways we produce, store, and use energy. An eventual shift from fossil fuels to other sources seems inevitable, while improvements in long-term energy storage and appliance efficiency are occurring rapidly. From the viewpoint of the energy consumer, however, little may seem to change. When an electric customer flips on a light switch, it makes little difference how the photons are produced, or whether the distant electrical generator is powered by solar radiation, burning coal, or nuclear fusion.

The Ultimate Energy Crisis

A thousand years from now, all the easily mined sources of fuel will have been exhausted. Near-surface layers of coal and concentrations of petroleum and natural gas will have long since been burned away. All the rich uranium mines, similarly, will have been abandoned. Sophisticated solar energy technologies and fusion reactors may by then have solved all society's energy needs. Or perhaps the great energy corporations will respond with trillion-dollar high-tech extraction plants that process millions of tons of rock and soil daily in the quest for atoms of uranium and carbon. The day will likely come when technology has advanced so far that the question "Will we run out of energy?" is no longer asked.

But this future world, totally dependent on high-tech solutions to energy use, will be poised on a dangerous precipice. Advanced technologies have evolved only once on planet Earth. That evolution took place in a world of abundant natural resources—seemingly inexhaustible supplies of coal and oil, and rich concentrations of metal ores that could be plucked off the surface of the ground and smelted in an open fire. In such a world of abundance, the discovery of metal technologies and exploitation of fossil fuels was perhaps inevitable.

If society ever suffers a relapse to a dark age—if technological know-how is lost through global war, epidemic disease, or the catastrophic impact of an asteroid—it would be far more difficult for humans to regain our present level of technology. A worldwide energy grid, powered by fusion reactors and solar collectors, depends on a large and highly skilled workforce. Once damaged, such an infrastructure could not be easily restored.

Without a reliable energy network, society would collapse. Cities would become death zones with soaring skyscrapers as lifeless as Egyptian tombs—without lights, water, or ventilation, and stripped of everything that can burn. It could come to pass in such desperate times that little of what we now know as civilization would be left. Libraries of ancient books and manuscripts, museums of combustible paintings in combustible frames, wooden altars and pews of the churches may all be consumed by energy-starved mobs.

Fossil fuels and metal ores are the bootstrap by which a society becomes technological. Exhausted of these resources, Earth may have to wait hundreds of millions of years while new supplies are formed.

CORE KNOWLEDGE: WHAT'S GOING ON INSIDE THE EARTH?

> Neither you nor anybody else knows for certain what's going on inside the Earth.
>
> —Jules Verne, *Journey to the Center of the Earth* (1864)

Inner Limits

A deep abandoned coal mine on the Japanese island of Hokkaido is the site of one of the world's most remarkable laboratories. There, a bullet-shaped capsule plummets straight down a half-mile-long vertical shaft, providing researchers with a precious ten seconds of free fall in which to do microgravity experiments. The vacuum-sealed capsule, which can carry up to a ton of experimental hardware, is released at the surface and drops vertically more than 1600 feet before a cushion of air slows and stops the descent. Then the cylinder is raised back to the surface for another free-fall experiment.

Eventually, if sufficient interest and research money are forthcoming, a similar facility might be lengthened, perhaps to as much as two miles, to give about twenty seconds of microgravity. But to go much deeper, where temperatures and pressures soar, is fraught with difficulty and danger. No known materials can withstand the rigors of the Earth's deep interior.

It is intriguing to imagine what it would be like to traverse a free-fall vacuum tunnel that extended through the entire Earth. It would take slightly more than a minute to fall 15 miles through the crust into the upper mantle, where rocks glow a dull red. Every second, speed would increase by another 22 miles per hour; in just six minutes a falling object would pass into the lower mantle, 400 miles down at a speed of almost 8000 miles per hour. It would take 13 minutes to descend 1800 miles to the boundary between the Earth's rocky mantle and metallic core, where the temperature is more than 3000° C and pressure exceeds a million atmospheres. And in about 20 minutes, traveling more than 20,000 miles per hour, a free-falling object would pass through the very center of the Earth, where conditions reach 3.5 million atmospheres and 6500° C temperature. With no resistance to slow its passage, the speeding cylinder would complete its trans-Earth passage in another 20 minutes or so, its deceleration mirroring the acceleration of the first 20 minutes.

Such an experiment will never happen. Our imaginations know no bounds, yet the laws of nature do. Science fiction writers explore the intriguing possibilities of teleportation, time travel, and faster-than-light voyages, but these technologies will remain fiction unless radical new physical laws are discovered. Based on our understanding of the universe, there are some feats that we can never accomplish, some places that we can never visit, and some questions that we can never answer.

Earth scientists study our planet under these daunting limitations. Based on what we know about the strengths of materials and the life-threatening effects of extreme temperature and pressure, we cannot conceive of a way to journey to the white-hot, super-pressurized center of the Earth. For a human to descend even a few miles into the crust—a tiny fraction of the planet's radius—is beyond any plausible technology.

Nevertheless, the Earth is our home and we are driven to know how it works. Questions about the structure of the cosmos, the fate of the universe, and the fundamental building blocks of matter fire our imaginations and invite deep philosophical speculation. But these mysteries are far removed from the concerns of everyday life. The science of the Earth, by contrast, may seem prosaically grounded in rock and soil. But studies of our planet take us on fantastic voyages in time and space. Our planet's dynamic past, present, and future are governed by epic forces acting over immense spans of time. A deeper understanding of powerful Earth processes, furthermore, may help us to avert some of the most devastating natural disasters.

THE PLATE TECTONICS REVOLUTION

New Questions

The Earth's crust, the thin accessible outer layer of our planet, provides all the raw materials we use in our daily lives, from metals, petroleum products, sand, and gravel, to skin and bones. The crust also bears the brunt of our planet's violent moods. Earthquakes and volcanoes, hurricanes and tornadoes, floods and droughts, are inevitably tied to surface processes. A few decades ago, geologists focused almost exclusively on these surface phenomena. Questions about the remote inner Earth, in comparison, seemed esoteric and removed from day-to-day practical concerns.

Today, geologists see the Earth's deep interior as a key to understanding every facet of our planet's past, present, and future. This new awakening to the importance of the Earth's interior arose in the 1960s with the discovery of plate tectonics, a revolution in the earth sciences as profound as any in its history. The crust is not isolated from the Earth's interior. Rather, the complex behavior of the thin surface layer is controlled by ceaseless motions of material deep inside the planet.

White-hot rocks a thousand miles beneath our feet, easily deformed by forces of gravity and buoyancy, flow in great currents that span much of the planet's interior. Over millions of years, hotter rocks near the core rise to the surface, while cooler rocks near the crust sink downward, much like the rolling action of a boiling pot of water. These epic rock cycles, called convection cells, push the relatively thin and brittle material near the surface to and fro, almost as an afterthought.

Ceaseless mantle motions have fractured the Earth's brittle surface into about a dozen plates, each sitting atop a convection cell. These rigid slabs range in size from hundreds to thousands of miles across, though they are no more than a few tens of miles thick. Each plate shifts across the Earth's surface in response to mantle flow, just as an oil slick follows the motions of the water on which it floats. Most plates are covered by oceans, but a few plates carry jumbled piles of lighter rock—the continents that we call home.

As plates are moved about, violent events occur at the Earth's surface. When a plate is split in two and pulled apart at a divergent boundary, new rock must rise from the mantle to fill the gap. Chains of volcanoes, such as those found along the Mid-Atlantic Ridge and the Great Rift Valley in Africa, mark places where new plate material is being formed at such a boundary.

When two plates collide at a convergent boundary, one plate must be swallowed up by the Earth. As a cool, brittle plate plunges down into the hot mantle, it may bend and crack, causing deep earthquakes. A descending plate may also partially melt, producing a long chain of violent volcanoes. Indeed, most of the earthquakes and volcanoes that affect the American Northwest, the Andes, and islands of the Pacific Rim occur at such convergent plate boundaries.

And when two plates scrape past each other, as they do along California's infamous San Andreas Fault, periodic violent earthquakes will ensue. Over time, roads, streams, and other features that cross the fault are displaced laterally as southern California moves, on average, two inches per year to the northwest relative to the continental United States.

What's Inside Affects What's Outside

Colliding plates shape the surface of the globe, creating supercontinents sutured by giant mountain ranges of crumpled rock along the line of plate impact. India and China, once continents on separate plates, have smashed together to form the mighty Himalayan mountains. The very summit of Mount Everest is capped by limestone—sediments originally deposited as an ocean reef. Hundreds of millions of years ago, a similar collision between the continents of Europe and North America created the Appalachians, which were once the mightiest mountains in the world.

Other continents are ripped apart to create new oceans. The Atlantic Ocean began to form a mere 200 million years ago—a small fraction of geological time—and it continues to grow wider as Europe and the Americas diverge by about an inch per year. Tens of millions of years from now, a new sea may divide a slice of western Africa from the rest of that continent.

Plate tectonics theory tells us that the most dramatic and seemingly permanent features of the Earth's surface—oceans, continents, mountain chains, ice caps—change constantly in response to the dynamic behavior of the Earth's deep interior. Shifting continents and oceans change climate by altering the course of ocean currents and prevailing winds that distribute equatorial heat and precipitation across the globe. The history of life is also tied to this dynamic quality of the Earth. As migrating continents collided and broke apart, life-forms adapted to new environments and evolved to face new competition. Even the depth of the oceans, which dictates the shape and extent of coastlines, is affected by plate motions. The continent of Antarctica, for example, is now at the South Pole. The southern ice cap has grown in places to a thickness of more than two miles, thus lowering the depth of the world's oceans. At times in the distant past, however, both poles were covered by oceans; the polar ice cover was much thinner and continents were inundated by shallow inland seas.

Plate motions also dictate the distribution of the Earth's resources. Stresses at plate margins lead to folding and faulting of rock, which provides pathways for the fluids that form rich ore deposits such as those found in the mountainous American West. The stable interiors of plates, on the other hand, often feature broad, fertile plains, such as those of the American Midwest. We can now say, without exaggeration, that virtually every phenomenon of the Earth's surface reflects what is happening inside.

Is the Earth Unique?

Scientists want to know what's going on inside the Earth because they are driven to understand our home. Is Earth an ordinary planet, just like millions of others in the galaxy, or is it (and, by extension, are we) somehow unique? To answer that question, we must learn what the Earth is made of—what rocks and minerals occur in the core and mantle—and we must identify the large-scale structures those materials form. In order to understand the dynamic interior, we have to discover if the planet's many layers are isolated from each other with smooth surfaces, like those of an onion, or if the layers interact with rough and irregular contacts. We ask what extreme temperatures and pressures exist beneath us to drive plate tectonics, and we wonder about the properties of Earth materials at those extremes. Most of all, we want to understand how the dynamic behavior of deeply buried rocks and minerals gives rise to earthquakes, volcanoes, and the other dramatic phenomena we experience at the Earth's surface.

What Is the Earth's Core Made Of?

"Earth's Center May Be Solid Mass of Precious Metals!" proclaimed a 1923 headline of *Popular Science*. The brilliant and eccentric geologist Henry S. Washington reasoned that the heaviest metals would sink to the Earth's center, forming a dense core with millions of tons of gold and platinum. Washington wasn't troubled by the lack of data to back up his theory. After all, without any evidence, no one could prove him wrong.

In Washington's day, speculating about the Earth's deep interior was a casual scientific pastime, a game played by gentlemen with vivid imaginations. Scientists of the day believed that our planet's deep interior was a static place, forever isolated from the outer layers of air, water, and life.

Today, geologists see the Earth's interior as a dynamic zone of epic forces and seething white-hot molten rock—a restless region that affects every aspect of the surface. The most destructive volcanoes and earthquakes are incidental burps and wiggles of the vast interior; the loftiest mountains and broadest oceans are mere afterthoughts of the Earth's ceaseless inner turmoil. Understanding what's going on inside the Earth has become *the* question for a new generation of geologists. At present, we have no way of knowing whether the Earth has a small inner core of dense precious metals, but we're getting closer to an answer.

Observing the Core Indirectly

We know so little about the Earth's interior: We are better informed about the surfaces of the moon and Mars than about what is hidden inside our own planet. Since we can't ever travel to the Earth's molten interior, we have had to invent plausible, imaginary descriptions of what the inner Earth might be like, based on indirect evidence.

Scientists' efforts to devise a realistic picture of the structure and composition of the Earth's deep interior are called geophysical modeling. Modeling is a scientific game played with a few simple rules. We must use what we know about the Earth's interior to devise a consistent description of the planet's chemical composition, internal structure, and dynamic behavior.

A hodgepodge of data constrains geophysical models. From observations of the orbital velocities of the moon and artificial Earth satellites, we know the total mass of the Earth to be about six trillion-trillion kilograms. Using special thermometers buried in rock, we can measure heat flowing from the interior to the surface, and thus have some idea of the planet's total internal heat supply. Satellites enable us to make accurate measurements of the size and shape of the Earth, as well as the shape and strength of the magnetic field, which originates in the core.

Rock samples provide hints about the Earth's chemical composition. Occasionally, geologists find actual samples of rocks from as much as 150 miles deep (intriguing reports of rocks originating from depths of 400 miles or more are still under investigation). These chunks of the mantle, ripped out and carried to the surface in violent volcanic eruptions, tell us much about the composition of the mantle. Meteorites, ancient rocks that fall onto Earth from space, also provide clues about the material that must have accumulated 4.5 billion years ago to form the planets of the solar system. Those samples reveal that silicon, magnesium, iron, and oxygen account for more than 90 percent of the Earth's mass. But the exact composition, the inner distribution of chemical elements, and the detailed nature of the rocks deep inside the Earth remain mysteries.

A geophysical model, as with any scientific hypothesis, must make testable predictions about the planet, and must be subject to modification as new data become available. No model can be proven to be *the* correct one; every model risks being disproved with a single new unanticipated observation. New observations and experiments are the key to success, because each new piece of data adds more constraints and thus limits the number of plausible answers. Progress is made, not by finding *the* correct model, but by eliminating impossible models.

In the coming decades, two groups of researchers will gain new insights into the Earth's interior. Seismologists reveal hidden structures by studying the distinctive paths of sound waves traveling through the Earth, while mineral physicists attempt to duplicate the conditions of the interior in their laboratories.

How Do We Probe the Earth's Interior

Scientists must explore the inside of the Earth without ever going there. Each of us solves similar problems every time we go shopping for fresh fruit. If we want to know about the inside of a piece of fruit without slicing it open in the store, we squeeze a tomato or tap a watermelon to see if it is ripe. In essence, we probe the hidden insides of the fruit by applying energy, and detecting the fruit's response to that energy. If the tomato feels squeezable yet firm and the watermelon sounds resonant, chances are they are ripe.

To know the Earth's interior also requires that we probe it, not unlike tapping a melon. Geophysicists do this by measuring the behavior of sound waves, or seismic waves, produced by any small earthquake or explosion. Seismic waves provide an excellent planetary probe because their exact speed depends on the kind of material through which they pass. Furthermore, as these sound waves spread out from their source, they may be bent or reflected in distinctive ways by different layers of material.

This exploration is made a bit easier by the fact that every seismic event produces two different kinds of waves that behave somewhat differently and give us complementary information. Primary, or P waves pulse through the Earth at speeds approaching 20,000 miles per hour. These fast-moving signals are compressional waves that result whenever matter is suddenly pushed. Drop a book, bang the table, or hammer a nail, and you cause a compressional wave. Secondary, or S waves, which travel at about half the speed of P waves, are shear waves that occur when matter vibrates side to side. You can watch a shear wave in action when a taut guitar string is plucked. Unlike P waves, S waves are unable to travel very far through a liquid or gas.

Any time a P or S seismic wave passes a boundary between two different kinds of rock, the wave may be partially reflected like an echo, and partially transmitted, in much the same way that light may be both transmitted and reflected off the clear, calm surface of pond water. Around the globe, arrays of sensitive detectors called seismometers record the arrival times of seismic waves and their echoes. Seismic observations made early this century documented that the Earth is made up of concentric layers—a thin outer crust, a thick rocky mantle, and a dense metallic iron core. The disappearance of S waves as they pass near Earth's center, furthermore, suggests that a portion of the iron core is molten.

The speed of sound waves through portions of the interior also points indirectly to specific minerals and rocks. Seismic wave velocities are most closely tied to a rock's density (the faster the wave, the denser the rock). We may not know the exact minerals and their compositions deep within the Earth, but thanks to seismology, we have a pretty good notion of how rock density varies with depth. We know, for example, that the Earth's average density is about five times that of water, with a maximum of about twelve times water's density at the center.

How Do We Map the Inner Earth?

Mapping details of the Earth's inner structure is not unlike the problem faced by a neurosurgeon, who must map a patient's brain before operating, but who is unable to dissect the living organ. Locating a small, dense object like a BB poses the simplest problem. Two X rays taken from different angles are all it takes to pinpoint the location of the tiny metal sphere. But how can you document a large, irregular structure like a brain tumor that may be intergrown with healthy tissue? One solution is a computer-aided mapping procedure called tomography.

Different regions of the brain are highlighted because they absorb X rays to slightly different degrees; for example, X rays are more readily absorbed in regions dense with blood (including blood clots and tumors) than in normal tissue. While no single X-ray photograph can reveal the exact location and shape of a tumor, measurements of X-ray absorption through many different paths provide the computer with enough information to produce a detailed three-dimensional map of all brain structures. Computerized axial tomographs, or CAT scans, are now routine diagnostic tools in medical centers around the world.

Seismologists take advantage of a similar tomographic technique to map the structure of the Earth by using the fact that seismic waves travel at different speeds through different materials. As more and more data are collected by increasingly sophisticated arrays of seismometers, powerful computers are able to process the vast seismic data banks.

Millions of different paths combine to reveal three-dimensional details of how wave velocities vary deep inside the Earth.

Seismic tomography is particularly well suited to detecting irregularities within each of the Earth's layers. Tomographic analysis reveals broad regions of the mantle that are warmer and other regions that are cooler—temperature differences that drive mantle convection. In some places, where brittle tectonic plates plunge into the mantle at convergent boundaries, tomography suggests that the descending slabs may penetrate to depths of more than a thousand miles.

Tomographic observations raise new questions. Immense features deep in the Earth seem to persist for hundreds of millions of years in spite of the incessant mixing of mantle material. Geophysicists have found deep, ancient roots hundreds of miles thick under South America, and possibly under other continents as well. Yet, if the mantle constantly convects, how can any mantle structure survive for so long? This question at present remains unanswered. As more data are collected, and as the techniques of seismic tomography are improved, we will be given an ever more detailed picture of what's going on inside the Earth.

What Materials Form Earth's Deep Interior?

The dominant chemical elements in the Earth's interior—silicon, magnesium, iron, and oxygen—are the same as those found at the surface. But under the immense pressures exceeding three and a half million atmospheres at the center and white-hot temperatures exceeding the surface temperature of the sun, these ordinary elements are transformed into new, dense, and unfamiliar patterns.

Early in the twentieth century, the effort to study matter at high pressures and temperatures was driven in large part by the desire to make diamonds, which require 50,000 atmospheres of pressure (about 350 tons per square inch) at 2000 degrees Fahrenheit to form. In 1954, a team of scientists at General Electric's research laboratory in Schenectady, New York, developed a reliable diamond-making procedure in which they squeezed carbon between the hardened anvils of a giant press while heating it with a powerful electrical current. That method is now used to manufacture more than 100 tons of diamond abrasives per year.

The high-pressure technique gave geologists a way to mimic the Earth's interior in a controlled laboratory setting. By the 1960s, dozens of earth-science facilities could duplicate mantle conditions using large presses. Even higher pressures, exceeding the 3.5 million atmospheres estimated to occur at the center of the Earth, were achieved using a miniaturized device with diamond anvils. Pressurized samples in such a diamond cell, heated with a laser to thousands of degrees, may change in remarkable ways. Some samples compress to half their ordinary volumes and display properties more like metal than stone.

Geophysicists now rely on these studies, collectively referred to as mineral physics, to reveal the behavior of rocks and minerals deep within the Earth. At any given combination of temperature, pressure, and composition, rocks have well-defined density, P- and S-wave velocities, and other characteristics. Too much iron, or too little silicon, and the wave velocities measured by seismologists won't match the properties determined by mineral physicists. Mineral physics data may never reveal the mantle's exact composition; for any given seismic signature, many possible combinations of elements might provide a match. But mineral physics data can place severe constraints on what is possible.

Discovering Mountains at the Core-Mantle Boundary

Many of the most exciting advances in mineral physics involve the discovery of new deep-Earth minerals. Researchers heat and squeeze dozens of promising combinations of elements in the hope of finding something different, but only a few scientists have succeeded in creating something really new. Kathleen Kingma joined that elite group in 1994, and in the process shed light on one of the great puzzles of geophysics.

Seismologists have found clear evidence that the boundary between the Earth's rocky mantle and metallic core, almost 2000 miles down, is surprisingly rough and irregular, not smooth as might be expected for the contact between layers of two materials as different as oil and water. Huge mountains perhaps 50 to 100 miles high seem to float on the iron-nickel outer core. These immense features may disrupt the Earth's magnetic field and cause significant variations in the apparent position of the magnetic north pole. What forms these mega-mountains? Mineral physics offers the best chance of finding out by identifying dense high-pressure minerals that might sink through the lower mantle, to float on the molten metallic outer core.

Bright, energetic, and driven by curiosity, Kathleen Kingma was a top postdoctoral fellow candidate when she joined the high-pressure research group at the Carnegie Institution of Washington's Geophysical Laboratory in 1992. She focused on one of the "big questions," in the field: What happens when you pressurize silicon dioxide, the common mineral quartz that forms beach sand (and many of the popular "power" crystals in novelty shops)?

Three decades earlier, then Russian postdoc Sergei Stishov had used a giant press at Moscow's Institute for High Pressure Research to discover a dense high-pressure form of quartz at about 100,000 atmospheres. In ordinary quartz, every silicon atom is surrounded by four oxygens in a rather open crystal structure. Stishov's new dense form featured a more efficient packing of six oxygen atoms around each silicon. That new mineral, dubbed stishovite by mineralogists, immediately became a prime candidate for an important constituent of the Earth's mantle.

In the late 1980s, a Tokyo-based research team had observed possible signs of an even higher pressure transition, and in 1992 Geophysical Laboratory physicist Ronald Cohen used advanced theory and a supercomputer to predict that a new form of silicon dioxide should appear at about 500,000 atmospheres. In 1993, Kingma decided to test the predictions as she employed a diamond cell to push quartz to more than half a million atmospheres.

The work required meticulous preparations. Valuable diamond anvils had to be precisely aligned or they would shatter under the extreme pressure. The microscopic silicon dioxide sample, no larger than a speck of dust, had to be mounted with painstaking care. Her diamond cell had to be filled with high-pressure gas, in a process that could fail explosively. But the results justified the difficult effort. Kingma found that stishovite does indeed transform to an even denser form of silicon dioxide at pressures characteristic of the lower mantle. This dense new form of silicon dioxide, perhaps one of the most abundant minerals in our planet, also provided a possible answer to the puzzle of mountains at the core-mantle boundary. Those gigantic structures may be silica—giant piles of pressurized sand.

Perhaps the most surprising result of the past quarter century of mineral physics research is that fewer than a dozen major minerals now appear to account for almost all of the deep Earth's high-pressure environment. Experiment after experiment yields the same minerals. In many ways the materials of the Earth's interior appear to be far simpler than those of the chaotic crust, which contains thousands of different minerals, and where scientists are discovering dozens more each year.

CAN WE PREDICT NATURAL DISASTERS

Many of the Earth's most destructive events, great earthquakes and volcanic eruptions, are minor blips compared to mantle processes that take place on a vast scale. Many earth scientists are now asking if our newfound understanding of the deep interior can help us to cope with these devastating natural disasters.

Earthquakes

Today, no one can say when the next major earthquake will strike a given place. The stresses of plate motions build gradually, but they are released suddenly. This situation is something like bending a wooden pencil and predicting when it's going to snap. As you slowly increase the bending force on the pencil, you know it must break. But when?

Many seismologists are cautiously optimistic that a reliable earthquake warning system can be developed, though practical prediction may not be available for many decades. Puzzling premonitory events provide critical clues to earthquake early warning. Animals have been reported to become agitated in the minutes before an earthquake, as they seem to sense an impending shock. The electrical and magnetic properties of rocks near faults may change in subtle ways prior to rupturing. And some big earthquakes are preceded by many smaller ones, though no obvious pattern has been identified.

The most promising premonitory observations relate to the behavior of geysers like Old Faithful. Geysers occur when groundwater boils after coming into contact with hot rocks. Many geysers display predictable patterns of fountaining—you can set your watch by some of them. But a day or two before a big earthquake, some geysers have been observed to become more erratic. The husband and wife team of Paul Silver and Nathalie Valette Silver, examining detailed records for the Calistoga Geyser in northern California, discovered that it changed its pattern of fountaining dramatically from one to three days before three different large earthquakes, including the destructive Loma Prieta shock of October 1989. Most astonishing, all three quakes were more than 100 miles away from the erratic geyser.

Irregular geysers, increased minor earthquakes, and even agitated animals may stem from the same physical phenomenon. In the days before a "big one," stress must build up in rocks throughout a wide zone around a fault. Such stress may affect water levels, and thus the behavior of geysering, while causing subtle changes in rocks' magnetic and electrical fields.

Precursor events are necessary, but not sufficient, to predict earthquakes. A change in geysering pattern, for example, may signal an upcoming quake, but it provides no clue as to where. The key to earthquake prediction, therefore, may be to establish much more extensive networks of seismometers, geyser monitors, strain meters, and other detectors near known earthquake centers. Traditional seismic observations have been concentrated in a narrow strip along the most dangerous faults. A strategy targeting 100-mile-wide stress zones around active faults might provide a basis for more accurate prediction.

While earthquakes may be difficult to predict, some geologists suggest that we might eventually learn to control them. A possible simple strategy for earthquake control was discovered quite by accident in 1966, when the Department of Defense, in an effort to dispose safely of dangerous liquids used to make nerve gas, pumped fluids into a deep well. Much to their surprise, they observed a strong correlation between the quantity of liquids pumped into the well and earthquakes in nearby

Denver, Colorado. Apparently, the pressurized fluid lubricated a fault, causing it to slip.

By pumping fluids into a known dangerous fault zone, geologists might be able to induce sudden slippage, and thereafter keep a fault permanently lubricated. Communities like Los Angeles could designate an earthquake day—perhaps a mandatory vacation/evacuation three-day weekend during which stress along the San Andreas Fault is relieved. Some destruction would occur, but, all told, lives and property would be saved. Subsequently, the continuously lubricated fault might pose little danger.

Such experiments will not take place anytime in the near future, because scientists have almost no idea how earthquakes are triggered. Lubricating faults might cause a swarm of relatively safe minor quakes in one place, but might induce one epic giant quake in another. Until we know more about the mechanisms that store and release an earthquake's energy, we cannot even contemplate triggering one.

Are Volcanoes Our Greatest Geological Hazard?

Many volcanoes, including those that form the Hawaiian Islands, vent Earth's internal pressure gradually. Their basaltic magma flows steadily down slopes in a predictable path. Property may be lost, but most people easily escape the swath of destruction. Explosive volcanoes like Mount Saint Helens and Mount Rainier in Washington State are not so cooperative. Hot gases build up for months inside the bulging mountain like a tightly sealed pressure cooker. This awesome energy may be concentrated in a catastrophic explosion that releases a raging river of steam and incandescent pulverized rock. The July 1980 eruption of Mount Saint Helens was tiny as such events go, blasting out about half a cubic mile of rock, and destroying perhaps ten square miles of forest. But when Mount Jemez erupted in northern New Mexico about a million years ago, 200 cubic miles of rock devastated a vast area that extended as much as 300 miles to the northeast.

Mount Rainier, situated just 30 miles to the southeast of downtown Tacoma, is an active explosive volcano of similar destructive potential.

In 1994, the National Academy of Sciences issued a strongly worded report urging the close study of the now-quiet mountain that last erupted only 150 years ago. In response to the academy report, geologists now continually monitor Rainier's slopes with an array of seismometers and heat-sensitive infrared detectors. They know that the mountain will erupt again sooner or later. Meanwhile, the suburbs of Tacoma grow to the south and east, ever closer to the majestic sleeping giant.

CAN WE LEARN
HOW TO X-RAY THE EARTH?

Futurists imagine yet another way to know the Earth. First, we will need a source of energetic particles that can penetrate the Earth—particles that are partially absorbed and partially transmitted as they pass through the entire planet. We must then measure that energy like a medical X ray, but on the vastly greater scale of the Earth's interior. The problem is finding an energy source that can travel through the Earth, and designing a detector that can record that elusive energy.

X rays themselves are completely absorbed by a few inches of soil, so they clearly will not work as a probe. Neither will other forms of electromagnetic radiation: radio waves, microwaves, visible light, and so forth. Energy beams of electrons or neutrons wouldn't get through the first foot, much less 8000 miles, of rock.

But a beam of neutrinos could. Neutrinos—massless, electrically neutral particles discussed in Chapter 1—are produced in abundance by nuclear reactions in the sun, other stars, and distant galaxies. According to the standard model of neutrino formation, each second more than 100 billion neutrinos pass through every square inch of the Earth's surface.

These elusive particles come in a wide range of energies—a spectrum of neutrino energies. Most neutrinos that hit the Earth are lower energy particles from the sun; almost all of them pass completely through the Earth. But, according to widely accepted physical models, a significant fraction of the much more energetic neutrinos produced in the centers of distant galaxies should be absorbed during their passage through the Earth, and each isotope will preferentially absorb neutrinos of different energies.

If we could compare the energy spectrum of a beam of neutrinos before and after it passed through the Earth, we might be able to measure which neutrinos had been most strongly absorbed, and thus determine directly the composition of Earth's interior. In other words, by measuring thousands of neutrino energy profiles through different slices of the planet, we could produce a neutrino tomograph—a detailed three-dimensional map, not unlike a CAT scan, of the inner Earth.

While neutrino tomography is a distant dream at this time, a team of geophysicists, astronomers, and physicists at the Berkeley campus of the University of California has proposed the first steps toward this bold experiment. The Berkeley team's preliminary experiments will exploit Earth-based neutrino detectors called neutrino observatories by their designers. Several observatories, including ones off the Hawaiian coast, in the Mediterranean Sea, and in Antarctica, are now in the planning stages. These facilities consist of long arrays of detectors anchored deep in the ocean or buried in polar ice, to minimize interference with other kinds of radiation. Each absorbed neutrino produces a brief flash of light, which is recorded by the detectors. Eventually, perhaps late in the next century, neutrino detectors sent into orbit around the Earth could monitor extragalactic neutrino beams, both directly and in the Earth's shadow.

Daunting challenges must be met before we can harness neutrinos as a probe of the entire Earth. Astronomers must locate distant neutrino sources to be used as beacons of energy, like the focused X-ray beam in your doctor's office. Physicists must determine how neutrinos' energies are absorbed by different chemical elements. Engineers must design giant neutrino detectors to capture particles so elusive that they pass through the entire Earth. Finally, geophysicists must learn to interpret the novel neutrino data in their efforts to devise models of the Earth's interior.

Neutrino tomography may be a scientific long shot, but the effort is well worth the risks. Whether or not the project ultimately succeeds, we will learn much about our universe along the way, for fundamental questions about matter and energy lie at the heart of the effort. And, if someday whole-Earth neutrino tomography becomes a reality, earth scientists will have realized their ultimate dream—to peer into the very center of our planet.

THE FATE OF THE EARTH: HOW MANY PEOPLE CAN THE EARTH SUSTAIN?

We love to point fingers when we try to deal with difficult
problems such as the environment, to lay the blame on industry or
science or politicians. And there is no question that
industrialization has polluted our surroundings. But who buys the
products? We do, you and I.
—Jane Goodall, *National Geographic* (December 1995)

At the edge of chaos, unexpected outcomes occur. The risk to
survival is severe.
—Michael Crichton, *The Lost World* (1995)

The Vanishing Frogs

In the 1980s, naturalists began to notice a curious phenomenon. In
forests, fields, swamps, and lakes around the world from Borneo to
Boston, the evenings have become quieter. The frogs are disappearing.

The loss of a few amphibians may not seem to be all that important,
but some ecologists view the change with a sense of foreboding. Disappearing frogs are to these environmentalists what dead canaries were to
coal miners—a warning that something is desperately wrong in our
biosphere and immediate action is called for. Naturally, others, with
money and jobs at stake, see a much more benign explanation based on
natural population cycles or weather.

The trouble is, biologists really don't know why the frogs are disappearing, just as we don't know for sure about the changes that are
taking place to the planet. In the absence of a comprehensive theory of
environmental change, "scientific" conclusions reflect individual ideologies as much as objective truth. Some observers tend to think of the
environment as a fragile thing under assault by opportunistic industry,
uncaring governments, and an exploding world population. Others, believing the environment to be resilient and human needs of first priority, take a more optimistic position. Meanwhile, environmental scientists must grapple with an interconnected global system of air, water,
rock, and life of immense complexity—a system potentially on the edge
of chaos.

THE POPULATION EXPLOSION

At present growth rates, the human population will double to about 12 billion by 2025, and double again by mid-century. By these straightforward projections, the twenty-second century would see the population soar past half a trillion—a density of people comparable to that of metropolitan Los Angeles on every available square mile of Earth's land, including Siberia, the Sahara Desert, and Antarctica.

No matter what one's ideology or religious convictions, such a growth trend cannot be sustained. Farmland will become exhausted of soil and its nutrients, reducing food production; reservoirs of water will be drained, causing widespread shortages in urban communities. As humans take over more and more land, many of the planet's ecosystems will be crowded out and collapse. Looking into the future just a few generations from now, all other unanswered questions about our physical world may pale to insignificance beside the problem of overpopulation.

Recognition of this problem is not new. Two centuries ago, the influential British economic theorist Thomas Robert Malthus proposed that the population will inevitably expand faster than the food supply. In his sobering *Essay on the Principle of Population*, he painted a vision of a future of poverty and despair in which total extinction of the human species is prevented only by the companion miseries of disease, famine, and war. He argued that the improvement of humankind is not possible without severe restrictions on reproduction.

For two hundred years this doomsday scenario has been held at bay by technological advances in food production, transportation, energy, and communications. But no known technology can manufacture millions of additional square miles of habitable land, or bring back species of plants and animals once they are driven to extinction. Biology tells us that any ecosystem can hold only a limited number of individuals of a given species. There are absolute limits to the number of human beings the planet can support. Are we converging on a Malthusian catastrophe—a world in which desperate humans live miserable lives competing for dwindling resources?

Population experts find it fiendishly difficult to estimate the limits of a sustainable human population. The answer to this vital question rests on calculations that require untestable assumptions about such fuzzy factors as crop yields, changes in sea level, demographics, new technologies, and global weather patterns—far too many unconstrained variables to answer the question in any definitive way. In spite of the uncertainties, most current estimates of Earth's population limit range from less than five billion to about thirty billion people. The more optimistic view is that new agricultural technologies, careful management of resources, a global system to distribute food, and reduced consumption will support modest population growth for the better part of a century. Some naive commentators even suggest that by the time populations reach critical levels, we will have developed the means for large-scale colonization of space. (They do not explain who will pay for ships to transport billions of people into space, nor where these people will go once they get there.)

Experts who advocate the smaller limit of five billion people, however, fear that we are already overfarming, overfishing, and overbuilding. By their estimates, water, soil, fish, and other essential commodities are already being consumed faster than they can be replenished—a deficit spending of resources that will eventually lead to the collapse of the global environment. Whatever view one holds, optimist or pessimist, the rate of population growth clearly must be reduced, eventually to zero. Earth has only so much matter and energy to go around, and at some point we will tip the balance.

What Are the Consequences of Overpopulation?

Every living thing competes for two precious resources: atoms and the energy to shape those atoms into useful structures. The rich diversity of life on Earth reflects the incredible variety of strategies that have evolved to claim a share of Earth's finite bounty. As plants struggle upward to harvest the sun's radiant energy, they require water, minerals, and carbon dioxide as raw materials for the energy-rich molecules that form roots, stems, leaves, and fruits. All animals obtain their atoms and energy, either directly or indirectly, from plants. Throughout this process, atoms are used and reused countless times, while energy continuously flows from the sun, through the ecosystem, and eventually back into space.

As the population grows, humans alter the balance of matter and energy in several ways. First, we command an ever larger share of water and nutrients—resources essential to all living things. Natural processes gradually renew supplies of groundwater and nutrients in the soil, but human activities often consume these resources much faster than they are replenished. Gretchen Daily of Berkeley's Energy and Resources Group estimates that almost half the Earth's soil cover has been degraded to some extent. Similarly, groundwater reservoirs critical to agriculture in the American Midwest are being consumed much faster than they can be replenished. Each new human commands additional resources and places further demands on the Earth.

In addition, habitats are eliminated through deforestation, farming, urbanization, and other endeavors. By limiting the amount of food and energy available to other organisms, these activities can cause large-scale changes in local ecosystems and may affect the diversity of animal and plant species. During the past half century, for example, the United States has lost more than half its wetlands, while more than 90 percent of Brazil's vast tropical rain forests have been destroyed.

Humans also affect environments by eliminating species. Many large animals have been hunted to extinction, and hundreds more remain on the endangered species list. Today, in addition to the much publicized destruction of whale, tiger, rhinoceros, and other mega-fauna populations, high-tech overfishing of the world's oceans has devastated cod, haddock, and other fish stocks. It seems improbable to some commentators that a common species of fish could be eliminated, but a similar fate befell the passenger pigeon. These North American birds numbered in the billions early in the nineteenth century, but they (along with a rich variety of microscopic symbiots) were wiped out by sportsmen in a few decades of overzealous hunting.

Finally, human activities result in a wide variety of air, water, and soil pollution, which may degrade ecosystems on a regional or global scale. Twenty years ago, for example, Chesapeake Bay, the largest estuary in the United States, boasted abundant water plants and birds, and teamed with commercially valuable crabs, oysters, and fish. Today, those populations have been devastated by chemical runoff from farms and factories as well as overfishing.

Each of these human actions—competition for soil and water, loss of habitat, overharvesting, and pollution—is undertaken for short-term need or profit, at the risk of long-term environmental degradation. The principal concern of many scientists is that we cannot predict the ultimate consequences of such deterioration. To understand the effects of modern technology on the environment, we need more reliable data about the planet's land, water, air, and life, and we need better theoretical models that predict their interactions.

DATA AND MODELS

People often expect scientists to provide incontrovertible evidence that a particular action has or will cause a quantifiable reaction. Newton's third law of motion does work magnificently for billiard balls and space shuttles, but the intricately woven ecosystems of the Earth seldom lend themselves to such simple analysis.

Some environmental cause-and-effect relationships seem obvious. A massive volcanic eruption in the Philippines spews immense quantities of ash into the upper atmosphere, and for the next year or two sunsets are brilliant and global temperatures drop slightly. Sport fishermen introduce the large predatory Nile perch into Lake Victoria, and the population of tilapia, a smaller native fish, suddenly takes a nosedive. Nor does it seem entirely coincidental that the last mastodon disappeared in North America shortly after the first human hunters arrived on the scene.

Other environmental changes do not have such an unambiguous cause. It's difficult to know what combination of circumstances led to the unusually large number of Atlantic hurricanes during the summer of 1995. Severe weather is one predicted consequence of global warming, but no one can yet be sure if that is the cause. Nor is it obvious what is causing extensive bleaching of Caribbean corals or the global disappearance of frogs. Are these phenomena simply cyclical anomalies, or are they dire premonitions of more drastic changes to come? Environmental science is so new that gradual changes, cyclic or not, that take place over many decades may be almost impossible to spot among normal random fluctuations.

Unlike some types of laboratory research, environmental science often has no obvious controls. Solid-state chemists and molecular biologists are usually able to control variables one at a time—a process that facilitates zeroing in on cause and effect. With only one Earth to study, however, it may prove impossible to attribute most changes to an unambiguous human or natural influence.

Pressure to Come Up with Answers

Unanswered questions of environmental science, more than those of any other scientific field, become mired in troubling uncertainty and rancorous debate. On the technical side, these questions are immensely complex. Thousands of interconnected factors contribute to the quality of the air we breathe, the richness of life in the oceans, and the interactions of humans and other organisms on land. No one knows all the linkages, nor can anyone predict a linear pattern of causes and effects with absolute certainty. Every scientific statement about our physical surroundings must be hedged with phrases like "based on present data" or "assuming" thus and so. Any conclusion about the environment must be expressed with broad uncertainties that present an easy target for rebuttal.

On the human side, questions about the environment are not neutral. Jobs, homes, health, and quality of life are at stake. The ethics of what we owe future generations versus the rights of people alive today constantly come into play. Every environmental crisis seems to boil down to an issue of money and property: a golf course versus an endangered species of butterfly; a chemical plant versus air quality; jobs of fishermen or loggers versus the health of an ecosystem. Dutch chemist Paul Crutzen, cowinner of the 1995 Nobel prize for his work on ozone depletion, warns: "Models [of human and environmental interactions] will play an increasingly important role in future national and international policy, putting considerable requirements on keeping science and politics apart."

Every successful scientist becomes adept at inventing explanations—making up stories—to account for every possible observation. Is a new chemical solution blue? Well, it's obviously because the solution contains divalent copper. Had the solution been red or any other color, an equally plausible explanation would be forthcoming.

I need look no further than my own research for choice examples of creative storytelling. After fifteen years of experiments studying minerals at high pressure, I had developed convincing explanations for why the Earth's iron-rich minerals tend to be more compressible than magnesium-rich varieties. Over and over this trend held true, in such common minerals as micas, olivines, garnets, and oxides. However, when I squeezed iron and magnesium silicate spinel, one of the Earth's most abundant mantle minerals, the exact opposite occurred. The magnesium spinel was *more* compressible than the iron variety. I was surprised, but the experiments were unambiguous. In spite of years of accumulated evidence and compelling models of why iron should be more compressible, it took only a few minutes to develop a plausible explanation for this anomaly. I rushed an article off to *Science*, where it was quickly accepted and garnered the cover photo. In other words, even though the results seemed to contradict years of study, my colleagues and I were delighted by the discovery, and we had no problem finding a loophole in our models to explain the unanticipated result.

Of course, cause-and-effect relationships play a central role in the physical world, and science has become extraordinarily successful at discovering them. That's why we can build skyscrapers that don't fall down and develop medical procedures that save hundreds of thousands of lives. But we are still far from understanding all the details of even the simplest minerals, much less complex ecosystems. Scientists, who can so easily explain away any unexpected result, can just as easily argue either side of a controversial environmental prediction. No scientist can hope to make predictions about the global environment with absolute confidence.

So, what do we do? Environmental models cannot provide exact answers. But just as analysts can predict fluctuations in the weather or the stock market within reasonable limits, environmental scientists can point to a reasonable range of likely outcomes. All they need is reliable environmental data to get started.

The Quest for Facts

President George Bush pooh-poohed the environmental crisis. In spite of his skepticism, or perhaps in light of it, in 1989, he instituted the multibillion-dollar United States Global Change Research Program, a monumental effort to monitor global change and guide environmental policies into the next century.

The centerpiece of this Mission to Planet Earth, as it has come to be known, will be the Earth Observing System, an array of satellites to be launched, beginning in 1998, designed to collect continuous data on land, air, and oceans. These orbiters, coupled with Canadian, European, and Japanese environmental satellites, will measure cloud cover, polar ice, vegetation, sea level, winds, ocean currents, global temperatures, chemical pollutants, and many other vital statistics. Scientists are optimistic that these data will provide an accurate base line on the Earth's present state, as well as a means to document environmental changes while they happen. Nevertheless, it may take decades before any definitive trends emerge.

Of one thing we can be certain: The climate will change. A growing body of fascinating paleoclimate research documents dramatic swings on time frames ranging from hundreds to hundreds of thousands of years. Attempts to reconstruct historic global climate change draw on many disciplines, including biology, geology, and chemistry. Traditional approaches begun in the nineteenth century rely on distinctive fossils and geological features. The discovery of ancient fossil palm fronds at Spitzbergen and in Antarctica, extreme northern and southern latitudes, provided some of the first indications that global climates have changed. Dramatic evidence of repeated regional glaciation during the past several hundred thousand years in Europe and North America suggest more recent periods of cyclical global cooling and warming.

Detailed investigations of the past thousand years point to the remarkable speed with which climate can change. Pollen and tree ring records reveal that about a thousand years ago the American Southwest was much cooler and wetter, and it supported extensive agriculture. Meanwhile, England was sunny and much warmer than today.

Much recent paleoclimate research has focused attention on thick ice accumulations that preserve atmospheric and climate data. Glacial ice cores more than a mile and a half long from localities in Greenland and Antarctica preserve a quarter of a million years of deposition. These cores represent an extraordinary continuous record of precipitation, prevailing winds, and other changes in climatic conditions. In addition, gases preserved in small air bubbles in the ice provide information on atmospheric concentrations of carbon dioxide, methane, and other gases.

Of special importance are new data on average annual temperatures, based on measurements of oxygen and hydrogen isotopes. Ice is solid H_2O that contains small amounts of the isotopes deuterium (hydrogen with one neutron in its nucleus) and oxygen-18 (two neutrons heavier than normal oxygen-16). The concentrations of deuterium and oxygen-18 incorporated into glacial ice are quite sensitive to temperature; each one-degree Celsius drop in average temperature results in a measurable drop in the amounts of these two heavy isotopes. By measuring isotopes throughout the entire ice core, paleoclimatologists have found that average global temperatures have varied over a range of about six degrees Celsius. Preliminary evidence suggests a clear pattern of glacial advance during cooler periods, followed by slightly warmer interglacial periods (including the present) when glaciers retreat. These ongoing studies are proceeding in earnest, for they may represent our best hope for understanding the dynamics of global climate change.

What Kind of Experiments Can We Perform?

Large-scale global change studies based on satellite and paleoclimate data are complemented by more intimate research on individual ecosystems. The usual protocol is to isolate a small system, such as a lake, a stream, or small patch of land. Researchers then remove one species or alter one variable for a long enough time period to compare any changes to those on a similar, unperturbed system that serves as a control.

In a now-classic 1961 ecological study of the rocky Scottish coast, biologist Joseph Connell manipulated the distribution of the barnacle *Balanus* and observed complementary changes in another barnacle,

Chthamalus. As simple as this experiment was in concept, the idea of establishing a clear cause-and-effect relationship between two species in the wild was revolutionary. In 1966, American biologist Robert Paine followed Connell's lead and induced dramatic changes in an ecosystem off the Washington coast by removing a species of starfish, the dominant carnivore, from one area. Within a year the common California mussel had overwhelmed that ecosystem, crowding out most other invertebrates and greatly reducing the diversity of local species.

These pioneering efforts guide many of today's ecologists, who modify ecosystems one variable at a time to see how they are affected. Researchers have changed the acidity of entire lakes, added iron and other nutrients to soils, and introduced or removed species in enclosed areas. Some plant or animal species can be added or removed and nothing much happens, while changes in other species may cause drastic alteration of the ecological balance.

The simple strategy of altering one variable of a complex system at a time pervades modern biological research, from genetics to brain science. This neat and clean surgical approach reminds me of my youthful attempts to understand my family's first black and white TV, which my father (an electrical engineer) built from scratch in the late 1940s. Each time something went wrong—the horizontal control or the brightness, for example—he'd rummage around in the back, replacing one glass vacuum tube after another until it worked perfectly again. As a small child, I began to think that each tube was in some way responsible for one property of the television; by removing one tube at a time, it seemed possible to understand how the complex box really worked.

Unfortunately, neither televisions nor ecosystems are quite that simple. All the components are complexly interconnected. Controlled experiments hold the promise of revealing general principles about ecosystem behavior, and they certainly point out the severe unintended consequences that can result from addition or removal of a single species. But these experiments cannot reveal the intricate interplay of many factors working simultaneously. Complex ecosystems, like televisions, may behave chaotically when something goes wrong.

Predicting Ecosystems by Computer

In the contemporary environmental scene, modelers are in the best position to make global climate predictions, and hence they serve as the final arbiters in many questions about human population and the health of ecosystems. Global circulation models, or GCMs, divide the Earth's surface into uniform blocks, typically a few hundred miles on a side. Each of these vertical blocks is further divided into a dozen or so horizontal slices containing land, ocean, or atmosphere. These elaborate schemes include detailed data for every box on ocean currents, wind speed and direction, humidity, distribution of mountains, cloud cover, vegetation, atmospheric composition, and temperature, among other variables. The sun's energy feeds into and out of the system, and ongoing exchanges of matter and energy between adjacent boxes are governed by laws of motion and thermodynamics.

In a typical experiment, the modeler sets up initial conditions and lets the model run through many cycles, watching the global climate change. The great advantage of environmental models is that scientists can run controlled "experiments." It's possible, for example, to shift the positions of continents to their ancient configurations to test the development of paleoclimates, or to alter cloud cover or change the concentration of carbon dioxide. Each change in initial conditions yields a different global outcome. Models of ever-increasing sophistication are now able to duplicate some details of historic climate change, and they make plausible predictions about such key issues as global warming, sea level change, and shifts in precipitation patterns.

In spite of the success of the GCM approach, no model is yet sufficiently complex to mimic the Earth in detail. The boxes are too large, and attempts to incorporate cloud cover and ocean currents and temperature variations are notoriously difficult. No matter how sophisticated the model, uncertainties will remain. Perhaps the greatest of these uncertainties result from the chaotic nature of global processes.

Chaos has achieved the lofty status of a cliché in conversations about the weather. According to chaos theory, the future of many complex systems, including the stock market, interdependent populations of animals and plants, and the changeable atmosphere and oceans, may be influenced in essentially unpredictable ways by minor perturbations today. One small investor's sale might alter the complexion of the entire stock market a few months down the road. The death of one individual in a large population of insects or plants might ultimately alter an entire ecosystem. And it is conceivable, at least in theory, that whether it's raining or sunny a year from today could be determined by whether you breathe out your next breath to the left or to the right.

Such a view of unpredictable chaotic systems, likely to swing to this extreme or that at the slightest whim, makes for great drama. In Michael Crichton's blockbuster *Jurassic Park*, as well as its best-selling sequel, *The Lost World*, the author paints a vivid picture of chaotic ecosystems delicately poised at the brink of self-destruction and collapse. Crichton's fictional alter ego, Ian Malcolm, pontificates on the hubris of humans, who believe that they can predict these violent swings of nature: "A predator dies off, and its prey grow unchecked. The ecosystem becomes unbalanced. More things go wrong. More species die. And suddenly it's over."

Humans often adversely affect both local and global environments, and Crichton's parable carries an important plea for restraint, but his extreme presentation of this point of view is flawed. For one thing, most minor perturbations of a system have minor effects—they get completely lost in the bigger picture. Look closely at the fancifully colored spirals and whorls of the beautiful and widely reproduced Mandelbrot sets, the icons of chaos, and you'll see that almost all of the surface is well behaved and regular. Only in very small and specific regions is complexity observed. Selecting a random point in the Mandelbrot set almost always places you in a well-behaved area. Similarly, it would take an extraordinarily unusual set of circumstances for a single small investor to alter the stock market, or a single breath the weather.

Any complex system that has been around for as long as the stock market, weather, or life must be self-regulatory to some significant degree. These systems may be unpredictable in detail, but they are unpredictable only within well-defined limits. If the Dow-Jones average sits close to 5000 points at opening bell, it will almost certainly close within a few percentage points of that value—it won't double, nor will it crash to nothing. The weather may be a few degrees warmer than usual this month, or a few degrees cooler, but it never snows during a Los Angeles summer, nor does it hit 90 degrees during a New England winter. Similarly, the species of plants and animals in an ecosystem will change with time, but life in one form or another has thrived for billions of years and has survived cataclysms of unimaginable destructive force.

Competitive complex systems tend to be self-correcting; they tend to evolve in an effort to survive and grow. Throughout the history of life on Earth, ecosystems have evolved as they adapt in response to environmental stress. Our real concern need not be the total collapse of the natural order, but, rather, the smaller-scale changes that alter environments—reducing food supplies, contaminating water, or polluting air—on the scale of human lifetimes.

That being said, complex systems can suffer catastrophes under just the right circumstances, if they are pushed too far. The Nile perch drastically, perhaps permanently, altered the Lake Victoria ecosystem. Coral reefs of the Caribbean are dying at an alarming rate, probably from a minor one- to two-degree increase in water temperature. And within the past few thousand years a modest global warming led to massive melting of polar ice and sea levels hundreds of feet higher than today. Improved global models represent the best hope of recognizing such devastating changes before they happen.

GLOBAL CRISES

The most pressing unanswered questions about the environment relate to specific human actions and their consequences. It's easy to spot many localized, short-term problems. A toxic chemical spill may kill everything in its path and authorities waste no time in requiring immediate corrective action. Almost every day newspapers report a new technology-related health hazard. We are bombarded by stories of chemical dumping, nuclear accidents, and oil spills. Reports link cancer deaths to artificial sweeteners, pesticides, and poor diet. Asbestos, radon, CFCs, and dozens of other concerns compete for public attention.

But as the problems become more gradual and less localized, and the details of cause-and-effect relationships become increasingly fuzzy, calls for action seem less compelling. Two well-publicized global problems, ozone depletion and the greenhouse effect, serve as examples. Both are the focus of intense scientific study and political debate.

Crisis over the Earth's Ozone

A half century from now, the salvation of the Earth's ozone layer may be viewed as one of the great environmental triumphs of our time, as well as an object lesson in the risks of new chemicals. In the 1950s, gases called chlorofluorocarbons, or CFCs, were found to provide cheap and nontoxic materials for refrigerants, aerosols, and other everyday applications. Production of freon and other CFCs quickly grew into a major chemical industry, with many benefits and no obvious downside.

Doubts about the safety of CFCs were first raised in the 1970s following a series of chemical experiments on ozone, a natural molecule composed of three oxygen atoms that represents a minor but critical component of the Earth's upper atmosphere. Ozone absorbs the sun's harmful UVB ultraviolet radiation, which can severely affect phytoplankton and plant growth and is a major factor in human skin cancer. The ozone layer thus serves as a shield for life at the Earth's surface.

In 1970, Dutch chemist Paul Crutzen demonstrated that ozone can be depleted by chemical reactions with nitrogen oxides, a common ingredient of smog. A few years later, Americans Mario Molina of M.I.T. and F. Sherwood Roland of the University of California at Irvine added to concerns with their discovery that CFCs can cause the rapid destruction of ozone through a series of simple chemical reactions. High in the atmosphere, normally stable CFC molecules are fragmented by ultraviolet light, which triggers the release of single, highly reactive chlorine atoms. If a chlorine atom (Cl) contacts an ozone molecule (O_3), the ozone will be split into two pieces, O_2 and ClO. Given enough chlorine, the ozone layer could be effectively destroyed by this reaction.

Chemists were at first optimistic that most chlorine atoms would eventually become locked into relatively stable chemical compounds. Those hopes were dashed in 1985 when a British scientific survey revealed severe depletion of ozone over Antarctica. Crutzen and others quickly realized that stratospheric ice particles, which concentrate over Antarctica during the dark months of winter, provide a surface on which chlorine compounds break down, re-releasing chlorine atoms. Given this rapid chemical recycling, a single chlorine atom can destroy huge numbers of ozone molecules. In subsequent years, detailed studies of Antarctic ozone concentrations have revealed dramatic year-by-year declines, closely correlated with stratospheric concentrations of CFCs.

In response to the growing danger, seventy nations signed the 1987 Montreal Protocol, a treaty to eliminate the production of CFCs by the year 2000. Congress voted to accelerate this schedule so that no CFCs will be produced in the United States after 1995. In recognition of crucial work that "contributed to our salvation from a global environmental problem that could have catastrophic consequences," the Royal Swedish Academy of Sciences awarded Crutzen, Molina, and Rowland the 1995 Nobel prize for chemistry. It was a bold and unusual selection, the first ever in environmental science and one that underscored the essential role of basic research in understanding how humans affect the planet.

Ironically, the very week that the Nobel prizes were announced, the United States Congress held new hearings to consider lifting the CFC ban. Conservative Republican representative Tom DeLay of Texas, who introduced legislation to lift the CFC ban completely, stated: "I am here today because I believe that the science underlying the ban on CFCs and the connection between health and ozone depletion is debatable." In spite of a decade of accumulating evidence and a general consensus among scientists around the world, DeLay managed to find two scientists willing to question the cause-and-effect link between CFCs and the ozone hole.

The thoroughly documented connection between increased CFCs and decreasing ozone has been accepted by the vast majority of scientists on the basis of a preponderance of the evidence. Might they be wrong? Yes. The ozone depletion might reflect a completely normal natural cycle. But do we want to take that chance for the short-term profit of a few chemical producers?

Is the Earth Becoming Warmer?

Global warming stands out as a starkly ominous feature on the landscape of environmental issues. The concept that increased concentrations of carbon dioxide and a few other atmospheric gases will create a greenhouse effect, thereby increasing global temperatures, has a simple and intuitive logic. Some prophets foretell of ecodevastation unparalleled in Earth history—an impending catastrophe of violent weather, collapse of agriculture, and coastal flooding that will push humans to the edge of extinction. Others fervently believe in a self-correcting planetary mechanism that will buffer whatever global indignities our species can devise. To their profound frustration, scientists just don't know the real answer.

Global-warming research advances on many fronts. The most basic data are ongoing measurements of global ocean and atmospheric temperatures, accumulated at the rate of millions of numbers every day by satellites and ground-based observers. Historically, most temperature measurements have been made in urban areas—regions that have experienced significant warming over the past century simply due to the increased area of pavement. Cities have become "heat islands." Detailed annual measurements of temperature in rural areas, on mountaintops, and over the ocean provide a better picture, and suggest that a small but statistically significant increase of about one degree Fahrenheit has occurred during the past half century. An international effort of one thousand experts endorsed this conclusion in the 1995 report of the United Nation's Intergovernmental Panel on Climate Change. Of great importance for ongoing documentation during the coming years will be infrared-sensing satellites that monitor the temperature of ocean water, which stores most of the Earth's near-surface heat. This Earth Observing System data will provide the most accurate baseline for evaluating future global change.

The other half of the global warming cause-and-effect equation is carbon dioxide. Decades of detailed measurements of atmospheric carbon dioxide provide unambiguous evidence that concentrations have increased dramatically during the past century, and by as much as fif-

teen percent in the last forty years alone, largely due to increased burning of fossil fuels.

Yet the correlation displayed by these ongoing measurements of
temperature and CO_2 concentration—each has risen in the past half
century—is not enough to prove cause and effect. It is conceivable, for
example, that the slight rise in temperature might be due to increased
solar energy output, which fluctuates in as yet poorly understood ways.
A short period of global cooling occurred during the "mini ice age,"
from about 1645 to 1715, when sunspots all but disappeared and the
sun is estimated to have been about one percent cooler. It is essential,
therefore, to document details of the complex global cycles of carbon
and oxygen. Where are the atoms stored, and how do they shift from
one reservoir to another?

The most compelling evidence for the greenhouse effect comes from
global circulation computer models, which invariably predict warming
associated with increased atmospheric carbon dioxide. Details vary according to cloud cover, vegetation growth, rate of polar ice melting,
carbon dioxide uptake by the oceans, and the mitigating effects of
sulfates and other aerosols, but every model predicts warming. In a
recent report summarizing more than 10,000 computer simulations, the
Environmental Protection Agency concludes that ocean levels will rise
by an additional half foot by 2050, and a foot by 2100, as a result of an
estimated four-degree-Fahrenheit increase in temperature by 2100.

In spite of such studies, some politicians from regions rich in fossil
fuels or those strongly dependent on their use discount the significance
of the greenhouse effect, calling global change research "scientific nonsense," at best "unproven, and at worst . . . claptrap." It is difficult to
respond to such extreme claims with purely scientific arguments. Uncertainties will always remain.

Where Science and Ethics Collide

We don't know the percent of ozone loss that is due to CFCs, but we
know it's not trivial. We don't know how much global warming will take
place during the next half century, but all models indicate it is taking
place. We can't predict the consequences of widespread species loss
and elimination of rain-forest ecosystems, but we do know such loss is

irreversible. In the face of such environmental projections, supported by the preponderance of evidence and endorsed by the vast majority of experts working in the field, it's only common sense to protect the environment. If we conserve, then the worst that can happen is that we consume a little less today and pass the Earth's resources on as a legacy to future generations, who may be better able to predict the consequences of their actions.

Scientific research on the environment is essential to address the magnitude of the problem. Biologists can seek new varieties of crops, geologists can devise more efficient uses of the land, and environmental scientists can model the complex interplay of economic, social, and physical factors that place limits on human survival. But the most troubling unanswered questions about the human population explosion and environmental change lie outside the domain of science. When we reach the inevitable realization that the population has grown too large, what do we do? Our fundamental belief that every couple has the right to have as many children as they want will come directly in conflict with the equally strong belief that every child deserves the resources to thrive. Ethical decisions regarding birth control, abortion, child welfare, foreign aid, and treatment of the aged and terminally ill may someday take place in a climate of fear for the survival of our species.

Of one thing we can be sure. Science alone cannot solve environmental problems, but without science these problems surely cannot be solved.

Origins: How Did Life on Earth Arise?

> Anyone who tells you that he or she knows how life started on the
> sere Earth . . . is a fool or a knave.
>
> —Stuart Kauffman, *At Home in the Universe* (1995)

Science in a Beaker

"Small" science can be important science. Such is the philosophy of
biologist Harold Morowitz who, armed with nothing more than a glass
beaker and a few simple chemicals, has been trying to duplicate key
steps in the origin of life on Earth.

"It's a $50 experiment," he claims. "Well, maybe $75."

In an era of multimillion-dollar grants, that's an astonishing admis-
sion, even more so considering that Morowitz is tackling one of the
oldest and most intractable questions in science.

The setting is a black-topped lab bench in a functional cinder-block
building on the George Mason University campus. As his chemicals mix
and react, Morowitz smiles broadly at the unorthodox simplicity of his
approach. "Of all origin questions, the origin of life is the easiest to
answer," he muses. "It is reducible to laboratory experiment—it's a
chemical process."

A self-described "philosophical plumber," Morowitz argues passionately that the chemical reactions that led to life must have taken place
in a low-tech environment. "Let's get it to go in the beaker first," he
concludes. "Maybe we'll need bolts of lightning. Maybe we'll need high
pressure. Maybe we'll just need a reducing atmosphere. But try the
beaker first."

HISTORICAL QUESTIONS

When researching a story, newspaper reporters are told to examine six
key questions: who, what, where, when, why, and how. Examining the
origin of life from the perspective of a news reporter reveals the intricate and often frustrating character of this grand old problem.

Although the origin of life was a historical event, many details of
that history have been irretrievably lost. The best we can now hope for

is to deduce and perhaps reproduce some of the chemical steps associated with that event. A review of the six key journalistic questions reveals why so much about life's beginnings remains beyond our ken.

Who Originated Life on Earth?

Ancient creation stories invoke a supernatural being as the first cause for the origin of life. "In the beginning, God created" is a theme common to many of the world's religions. Science can neither prove nor disprove such beliefs when they are based on faith, and thus the question "Who originated life on Earth?" usually lies outside scientific inquiry.

A curious exception is provided by the "panspermia" theory, the fascinating suggestion that the universe has been seeded with life, perhaps by an unknown intelligent alien species. This ancient concept was first advocated in modern times by the Swedish chemist Svante August Arrhenius, an early recipient of the Nobel prize, and its banner has been carried by such luminaries as astronomer Fred Hoyle and biologist Francis Crick. According to an extreme version of this idea called "directed panspermia," life on Earth is an outgrowth of a conscious effort to colonize the galaxy.

As strange as panspermia may sound, it is, at least in principle, a testable hypothesis that makes specific predictions. The theory predicts that such seeds exist and that life on other worlds, if derived from similar seeds, will bear striking similarities to Earth's. No one is likely to prove panspermia advocates wrong anytime soon, but the obvious shortcoming of this solution to the question of life's origin on Earth is that it merely displaces the great mystery of the first living cell to some more distant time and place. Some sequence of events must, once upon a time, have led from nonlife to life. And the suggestion that there was (or is) a conscious effort to colonize the galaxy speaks only to whether some living thing dispersed the seeds. In principle, the sower might be discovered by the search for extraterrestrial intelligence (see Chapter 14).

What Is Life?

In thinking about the origin of life, it is essential to define a set of minimum characteristics that distinguish living from nonliving objects. This task is not trivial; indeed, no single definition of life is accepted by the entire biological community. Rather than add to that controversy here, we merely identify a few key characteristics that are now shared by all life forms on Earth:

(1) All life consists of one or more cells, in which the chemical processes of life occur. Membranes, constructed primarily from a double layer of polar lipid molecules, form a barrier between the cell's water-based interior and its surroundings.

(2) All life uses carbon dioxide, directly or indirectly, to construct carbon-based molecular building blocks: primarily carbohydrates, proteins, lipids, and nucleic acids.

(3) All life obtains useful energy through the chemical reactions of metabolism, notably those involving the formation and breaking of bonds between phosphorus and oxygen, and oxidation-reduction reactions.

(4) All life stores the information required for growth and reproduction in the form of nucleic acids. The genetic code carries instructions for manufacturing the proteins that build cellular structures and facilitate specific chemical reactions in the cell.

This four-part description of life on Earth is pragmatic and not necessarily applicable to life that may exist anywhere else in the universe. Nevertheless, these four characteristics are of central importance to thinking about how life originated on our planet, because together they represent minimum structures and functions that must have been present in the first living cell.

Where Did Life Originate?

The Earth's continents are in constant motion, shunted about by the dynamic process of plate tectonics. As a result, the Earth of one hundred million years ago looked quite different from today, although the modern continents would probably have been vaguely recognizable on the jumbled globe. A half-billion years ago, however, the distribution of land and water was vastly different from today. And no one has the vaguest idea of what the Earth looked like four billion years ago, at the dawn of life, except perhaps for the familiar swirling of white clouds against the pleasing blue of the seas.

Given dramatic historic changes in the Earth's surface, the question "where" loses most meaning. No reference points defined the longitudes of the ancient place in which the first cell appeared. At best, if a likely life-forming process is eventually identified, scientists might be able to estimate the relative probability of life originating in the warmer equatorial regions versus the cooler ones near the poles, or the sunny surface of the sea versus deep subducting sediments. And although we can't identify a spot or a region for erecting a monument to the first cell, it does seem likely that life had a wet start.

When Did Life Originate?

The origin of life can be bracketed with some certainty to a span of a few hundred million years. The Earth began to form around 4.5 billion years ago by accumulation of material from the solar nebula, a vast, swirling, flattened cloud of gas, dust, and rock. As the planet grew, a constant rain of meteors and comets pulverized the surface; the energy of these projectiles brought surface temperatures to the melting point of rocks. The largest of these collisions spawned the moon, when a Mars-sized body slammed into the emerging planet, disrupting the Earth to its very core and flinging a moon-sized blob of molten mantle into orbit. Judging from the moon's pockmarked surface, with its giant craters dated as young as 3.8 billion years, the lethal bombardment continued for several hundred million years after that "big splash." The exterior of the Earth was an inferno of volcanic eruptions, spewing molten rock onto the surface and superheated gases into the atmosphere. No life-forms could have survived the cataclysmic primordial bombardment and turmoil, which by some estimates lasted until somewhat less than four billion years ago.

Surprisingly, clear signs of single-celled life occur soon after in the oldest surviving rocks on Earth, those from the approximately 3.8-billion-year-old Isua belt in Greenland. Thin, transparent slices of this rock reveal what appear to be microscopic cells, some of them in the process of cell division. Thus, life was firmly established on the young Earth 3.8 billion years ago, not so very long after the violence subsided.

While no one has seriously suggested that we could ever know the exact date of life's origins, there is subtle reason why such a question would, in any case, have an imprecise answer. Since the formation of the moon 4.5 billion years ago, the Earth's rotation has been slowing down, primarily because of tidal friction. This effect is too small to measure in human lifetimes with all but the most accurate atomic clocks, but over spans of hundreds of millions of years the consequences have been remarkable. At the time of life's origin, the rapidly rotating Earth would have needed a calendar with more than 500 days per year, with each day shorter than eighteen hours. Thus, for the question of when life originated, as for the question of where, our modern frames of reference simply don't apply.

WHY DID LIFE ARISE?

Questions about why the natural world is the way it is often lie outside the realm of science. Researchers have found no way to answer why the universe is composed of quarks and leptons, why it began in the big bang, or time flows in one direction. For now we must be content to describe these aspects of our surroundings, and accept that that's the way the universe is. In the case of life, however, an answer to why it arose might be forthcoming.

Practical experience and natural laws convince us that many events are bound to happen. Place a pan of saltwater in the hot sun and salt crystals will form as the water evaporates. Throw a rock into the air and it will fall back to Earth. Fertilize your lawn and the grass will grow

faster. We experience countless examples of such cause-and-effect behavior every day, and we try to plan our lives accordingly.

At the opposite extreme, some natural phenomena seem almost arbitrary, as if they were governed more by chance than by predictable
events. The exact sequence of occurrences that led to the formation of
the Earth's moon, Hurricane Andrew's destructive path across Florida,
or the evolution of the human species are unique, very unlikely ever to
be repeated exactly anywhere in the universe. No natural laws were
broken during these processes, but the final results depended on countless minor perturbations that would have been all but impossible to
predict. Primitive societies tend to ascribe such one-time events to
"God's will," but science tries to understand the natural causes.

Where in this spectrum of predictability is the origin of life? Was
life's beginning on Earth a unique event resulting from an almost infinitely improbable juxtaposition of molecules? Or was life, as many experts now argue, as inevitable as salt crystals on any sunny, wet planet?
Could several "life-lines" have begun, only to disappear, leaving just
one successful line.

According to Nobel prize–winning biochemist Christian de Duve,
three "isms" have clouded our attempts to understand this aspect of
the origin of life. "Vitalism" views life as imbued with some vital spirit,
fundamentally different from other matter. "Finalism" assumes that
biological processes are directed toward some final goal (presumably
us). "Creationism" claims a miraculous intervention in the appearance
of life. None of these three conceptual frameworks meets the criteria
for scientific understanding. Rather, de Duve states, a scientific approach "demands that every step in the origin and development of life
on Earth be explained in terms of its antecedent and immediate physi-
cal-chemical causes." In other words, he argues, a scientific explanation
of life must consider only natural processes that can be investigated by
observation, experiment, and theory.

Was Life Inevitable?

Two opposing camps argue about the inevitability of life. Some biologists, notably Jacques Monod, who wrote *Chance and Necessity* in 1971,
argue that life, while consistent with natural laws, is a unique and

extremely improbable event. They believe that life arose through a complex and random sequence of steps that could never be duplicated, let alone in human lifetimes in a laboratory environment. Given the vast size of the oceans and hundreds of millions of years to mix things about, the right juxtaposition of building blocks just happened to occur once only on Earth. This situation is something analogous to the familiar concept that given enough monkeys, enough typewriters, and enough time, one of the monkeys will write *Hamlet*. According to this view, the reason we happen to be on the unique planet with life is that we *have* to be if we're here to ask the question. If these scholars are correct and life arose through some odd and irreproducible concatenation of circumstance, then origin-of-life research lies beyond the realm of scientific questions, and science cannot explain in any meaningful way why life began. For most researchers, such a "facile recourse to chance," as Christian de Duve describes it, is neither satisfying nor scientific.

Nobel prize–winning chemist Linus Pauling spoke for most of the scientific community when he said, "The most unlikely thing about life is its being unlikely." Some scientists fervently believe that we can learn, and perhaps eventually reproduce, the step-by-step chemical details that led from a barren, primitive atmosphere and ocean to one that teamed with single cells. These researchers propose elegant experiments in attempts to reproduce specific chemical reactions that may have occurred in the Earth's early oceans—and anywhere else in the universe with Earthlike conditions—and been involved in the formation of life. For them, life is an inevitable combinatorial property of the elements carbon, hydrogen, nitrogen, oxygen, phosphorus, and sulfur.

A logical consequence of this view is that whatever sequence of events occurred on Earth has probably occurred on countless other planets as well. In *Vital Dust: Life As a Cosmic Imperative*, Christian de Duve's persuasive and accessible 1995 presentation of these ideas, he states: "Life is an obligatory manifestation of the combinatorial properties of matter. . . . [It] was bound to arise under the prevailing conditions, and it will arise similarly wherever and whenever the same conditions obtain."

The answer to the question of why life arose could eventually come from either of two sources. If laboratory experiments produce a cell from water and other simple molecules, or, if we discover that life is common on distant Earthlike planets, then we'll know that life is an intrinsic property of matter. Without such convincing evidence, however, the concept that life on an Earthlike planet is inevitable remains an unproven (and difficult to test) scientific hypothesis. Arguments such as those of Monod and de Duve rely as much on personal belief and intuition as they do on any pertinent observations. This is not to say that arguments for or against "life as a cosmic imperative" are invalid or improbable. Intuition, prejudice, common sense, and hunches play a vital role in science, just as they do in other human endeavors, but they do not constitute evidence.

This philosophical debate is strikingly reminiscent of the origin-of-the-universe controversy between Georges Lemaître and Arthur Eddington (described in Chapter 2), and it has similar theological overtones. The Catholic priest Lemaitre believed in a single creation event, and thus embraced evidence for the big bang. Arthur Eddington, on the other hand, was an avowed atheist who preferred an eternal, steady state universe. Similar beliefs may underlie some scientists' preference for unique versus multiple life creation events.

Finally, we can hope that these debates, which up to now have pitted very opposing views against one another, will evolve to discuss the equally plausible possibility that life, although not unique, is extremely uncommon. For now, such shades of gray are infrequently proposed, but often in science both sides of an argument turn out to be correct.

HOW DID LIFE ORIGINATE?

Studies of the fossil remains of ancient life, combined with research on the anatomy, cellular structure, genetics, and biochemistry of modern living things, have convinced the vast majority of biologists of two central facts about the development of life: A self-replicating cell arose on Earth from nonlife about four billion years ago, and life has been evolving from that cell by the process of natural selection ever since.

The second of these key ideas, evolution by natural selection, has become a central paradigm for the biological sciences (see Chapter 10). Evolution, the modification of life forms over time, is now an established scientific fact—as well documented a natural process as gravity. Scientists who investigate evolution concentrate on biological variation and the intricate mechanisms of natural selection.

The Scientific Problem

By contrast, the sequence of natural processes by which life first arose from nonlife remains among the most profound unsolved mysteries. In the words of Harold Morowitz, "They deal with events that may have occurred only once and have then so altered the character of the world as to erase many of the vestiges of their prehistory. . . . The study of beginnings is *a priori* more speculative and lies at the fringes of scientific thought." The key origin-of-life question, and the only one that is clearly in the mainstream of contemporary research capabilities, is "How did life originate?" How could the primitive hot mix of water and simple atoms and molecules yield a complex cell capable of self-replication? Any complete answer to this origin-of-life question, which is unlikely to come in the foreseeable future, will encompass a continuous series of chemical steps that occurred during the geologically narrow period from about 4.0 to 3.8 billion years. This problem is, in essence, an extremely complicated extension of materials science.

This field of inquiry is firmly entrenched in academia, with all the trappings of a thriving intellectual effort. The International Society for the Origin of Life has its own conferences and its own peer-reviewed journal, *Origins of Life*. In the United States this research is supported in large part by NASA, and a significant research effort is spearheaded by origin-of-life pioneer Stanley Miller and his group at La Jolla, California.

The central paradigm of origin-of-life researchers is that life may be viewed as a collective property of six elements—carbon, which forms the backbone of all "organic" molecules, plus hydrogen, nitrogen, oxygen, phosphorus, and sulfur (a useful mnemonic is CHNOPS). Two intertwined questions frame many studies of life's origins: By what chemical pathways did life on Earth arise? And are those pathways so likely that life will arise on any Earthlike world? On the one hand, origin questions of this sort are especially problematic. On the other hand, the origin of life can be approached in a straightforward, pragmatic manner: Unlike the big bang origin of the universe, we can try to reproduce conditions of the early Earth in the laboratory.

One of the most extreme proponents of this strategy is Stuart Kauffman of the Santa Fe Institute. In his self-professed "renegade" thesis, *At Home in the Universe*, he states: "I hope to persuade you that life is a natural property of complex chemical systems, that when the number of different kinds of molecules in a chemical soup passes a certain threshold, a self-sustaining network of reactions—an autocatalytic metabolism—will suddenly appear. . . . Stunning developments in molecular biology now make it possible to imagine actually creating these self-reproducing molecular systems—synthesized life. I believe that this will be accomplished within a decade or two."

Life in the Lab

Two complementary research strategies converge on origin-of-life questions. The first approach, exemplified by the famous Miller-Urey experiment of the 1950s, begins with simplicity. These experiments, which in essence work forward in time from the basic carbon compounds that existed 4.5 billion years ago, reveal that the primitive oceans must have become stocked rather quickly with a variety of relatively complex organic (i.e., carbon-backbone) molecules.

Biochemists approach the origin of life by examining the chemical mechanisms of the most primitive single cells now living on Earth. The smallest of these, called mycoplasma cells, are only about a ten-thousandth of an inch in diameter. These cells are the least complex lifeforms known; they depend on their environment to supply many kinds of organic nutrients, including amino acids and nucleic acids. Cyanobacteria, in contrast, are larger and more complex single cells, but they have the ability to survive and reproduce entirely from the most basic ingredients—carbon dioxide, nitrogen, and water, plus a few mineral nutrients. In searching for examples most akin to primitive cells, the structural simplicity of mycoplasma and the chemical simplicity of cyanobacteria can illuminate different aspects of early life.

The cellular structures and metabolic pathways by which cells extract energy from sugar, for example, are common to all life-forms, and must have been present in some fashion in the earliest cells. By paring down metabolism to its most basic chemical reactions, scientists hope to glimpse a plausible sequence of events that might have occurred spontaneously, before the first cell began to reproduce. Laboratory experiments with the appropriate mix of carbon-based compounds help researchers work backward in time toward presumed characteristics of the oldest known life-forms of 3.8 billion years ago.

What Was the Early Earth Like?

If scientists are to understand the origin of life, they must first know the composition and temperature of the Earth's early atmosphere and oceans. Experts agree that the atmosphere four-plus billion years ago was quite different from today's roughly four-to-one nitrogen-oxygen mix. Oxygen, a highly reactive and corrosive gas is a product of photosynthesis and was not a significant primordial component. Rather, hydrogen, methane, ammonia, water, carbon dioxide, hydrogen sulfide, and nitrogen have all been proposed as major atmospheric components, though the exact composition is a matter of much speculation and debate. Whatever the mix, a range of temperatures, from boiling waters near volcanic vents to frigid conditions in the tenuous high atmosphere, provided varied environments for organic synthesis to take place.

HOW THE FIRST LIVING CELL
MIGHT HAVE EVOLVED

Most origin-of-life researchers suspect that several key steps in the primordial crucible must have led to the first living cell: (1) the synthesis of relatively simple organic molecules; (2) the linking of those molecules into the larger structures of polymers; (3) the isolation of those molecules by a primitive cell membrane; and (4) self-replication. While the details, relative significance, and exact order of these steps are matters of intense debate, a summary of key steps provides a flavor of origins research today.

Step 1: Organic Synthesis

Perhaps the least problematic (and, not coincidentally, the most studied) step in life's origins is the appearance of a rich variety of simple organic molecules in the primitive ocean. Two distinct sources of these molecules may have enriched the prebiotic Earth.

University of Chicago professor Harold Urey and his graduate student Stanley Miller launched the field of prebiotic chemistry in 1952 by devising glassware that sent electric sparks though an atmosphere of methane, hydrogen, and ammonia, which circulated above warm water. Much to their surprise, in a matter of days the simple solution turned from colorless to pink to red to brown as a rich broth of organic molecules formed.

The significance of the Miller-Urey demonstration, which showed a rich variety of organic molecules easily arises from the simplest of building blocks, was bolstered by the discovery of similar molecules in interstellar dust clouds, comets, and even some meteorites that have undergone the rigors of a fiery passage through the atmosphere. Fly-by studies of Halley's Comet, for example, suggest that as much as a third of its mass consists of organic molecules, while interstellar dust and meteorite debris that constantly rain down on our planet are often loaded with these components. Recent studies of meteorites suggest that they may also have provided a source of phosphorus compounds that are critical to metabolism. In fact, some scholars argue that extraterrestrial material provided the bulk of organic molecules from which life arose.

Whether the source of simple organic molecules was terrestrial synthesis or extraterrestrial accumulation, these chemicals were an inevitable component of the early Earth. The earliest oceans and sediments may have grown increasingly concentrated in these organic molecules, for there was no life to gobble up the rich mix of sugars, amino acids, lipids, and other molecules.

Countless variations of the Miller-Urey experiment continue to fuel the debate over life's origins. This research proceeds with a strategy not unlike that of material scientists: Try lots of different mixtures of gases and water-based solutions, subject them to a variety of plausible conditions such as high temperature and pressure, and see what happens. Among the products of these varied experiments are all of the key building blocks of life.

Amino acids: These nitrogen-bearing molecules are the building blocks of all proteins, vital chemicals in every living thing. Proteins serve as important structural components, and they act as enzymes that facilitate chemical reactions in every cell. Although hundreds of different amino acids can be synthesized, only twenty different types form virtually all the proteins in all living things. This surprising selectivity provides one line of evidence that all life-forms on Earth are related by evolution to a common ancestral cell.

Nucleotides: Four distinctively shaped molecules called bases—adenine, cytosine, guanine, and thymine (abbreviated A, C, G, and T)—provide the "alphabet" for the genetic language common to all living things. In RNA, thymine (T) is replaced by uracil (U). These bases link to a sugar molecule, either ribose or deoxyribose, which is itself linked to a phosphate (PO_4), to form nucleotides. These nucleotides occur in a precisely ordered arrangement in DNA, the double-helix molecule that stores and duplicates genetic information, and in RNA, the single-strand molecule that interprets the genetic message (see Chapter 9).

Lipids: Lipid molecules include oils, fats, waxes, and other materials formed primarily from carbon and hydrogen. These substances do not mix with water and tend to clump together when present in watery liquids. An important class of these molecules, called phospholipids, form all cell membranes and thus are critical to understanding how the first cell formed. Phospholipids are long, skinny molecules and are distinctive in that one end bearing the phosphate is attracted to water (hydrophilic), while the other end, the lipid end, is repelled by water (hydrophobic). When phospholipids are concentrated in water, therefore, they spontaneously form a double layer, with hydrophobic ends facing each other, and hydrophilic ends facing outward. Such bilayers form by a process of self-assembly because of the inherent chemical properties of phospholipids; no external manipulation is required. If such a lipid bilayer curves around on itself, it can form a small sac, which isolates water and other chemicals inside from the outside—an essential feature of every living cell.

Carbohydrates: These molecules of carbon, hydrogen, and oxygen, including all sugars, play a vital role in the metabolism of every living thing. The most primitive cells manufacture carbohydrates from carbon dioxide and water, and they use these molecules to store energy and manufacture other cell materials.

Step 2: Building Polymers

Amino acids, carbohydrates, nucleotides, and other small molecules function as "monomers," which are the individual segments that link together to form the long chainlike structures called polymers. Proteins, DNA, cellulose, and many other polymers provide the essential structures of living things. Research into prebiotic chemistry hits a serious snag when it comes to transforming monomers into polymers.

Small organic molecules form easily, as Miller-Urey–style experiments have shown time and time again, but they usually do so in extremely low concentrations. If small molecules are to interact, the solutions need to be more concentrated so that two molecules will have a chance to bump into each other. The large and complex molecules essential to life—proteins with dozens or hundreds of amino acids, RNA and DNA formed with long chains of nucleotides, or cellulose assembled from countless individual sugar molecules—do not appear spontaneously, even in carefully devised environments with high concentrations of monomers. Indeed, these macromolecules appear to be quite unstable. Even when two monomers link up or polymerize, they often will just as quickly disintegrate or depolymerize under water-rich conditions.

Decades of intense and creative research have failed to identify any chemical mechanism by which large-scale molecules of life can be synthesized using Miller-Urey–style prebiotic processes. This is not to say that researchers won't keep trying. But it has long appeared that an alternative strategy, called catalysis, is required. In catalysis, an intermediary molecule (I) facilitates the linking of two monomers (A and B) into a polymer (AB), without I itself being changed. The rate of the polymerization reaction may thus increase by many orders of magnitude, overwhelming the tendency toward depolymerization.

Catalysis forms a central theme of Christian de Duve's thesis on the origin of life. He imagines an early atmosphere rich in hydrogen sulfide that facilitated the synthesis of a class of molecules called thioesters. According to de Duve, thioesters may have played a key catalytic role in producing large proteinlike molecules, which in turn served as catalysts for other critical reactions. These catalytic steps have the added appeal that they still play a major role in the metabolism of modern cells.

Alternatively, some biochemists have pointed to a possible role for catalysis by RNA, a chemical that helps to process DNA's genetic message in all living things. Many researchers were astonished in the mid-1980s when Sidney Altman at Yale and Thomas Cech and his colleagues at the University of Colorado at Boulder discovered that RNA molecules can act as catalysts in the manipulation of genetic information. Perhaps the first self-replicating entities were based on RNA chemistry rather than proteins. Still, a huge problem remains to be solved: How were the earliest RNA-like molecules formed? A variety of schemes can be imagined, but the experimental demonstration of that feasibility is needed.

Yet another intriguing option, advocated by Stuart Kauffman of the Santa Fe Institute, is "autocatalysis," a process by which molecules catalyze their own formation. In one of the simplest situations, molecule AB catalyzes the reaction of A plus B to form a new molecule BA; BA then catalyzes the reaction of A and B to form AB. A solution containing only molecules A and B may be stable for a long time, but add a bit of either AB or BA, and autocatalysis takes off in a closed chemical loop. This idea is nothing new, but Kauffman takes it a step further. He argues that as the number of different kinds of molecules in a system increases—a likely scenario for organic molecules in the prebiotic soup—then the probability of generating more and more complex autocatalytic networks rises sharply.

"Imagine a whole network of these self-propelling loops," he says. "If a sufficiently diverse mix of molecules accumulates somewhere, the chances that an autocatalytic system—a self-maintaining and self-reproducing metabolism—will spring forth becomes a near certainty. If so, then the emergence of life may have been much easier than we have supposed." In *At Home in the Universe*, Kauffman presents an accessible theoretical treatment of how such complex autocatalytic networks arise. Noticeably lacking from Kauffman's theory, however, are any suggestions regarding the chemical details of the process. In particular, it is not obvious how the requisite molecules could be present in sufficient concentrations to initiate any reactions at all.

Finally, catalysis need not be dependent entirely on organic molecules. Researchers have long known that the surfaces of inorganic materials can serve quite well, as they do in the catalytic converters of most automobiles. But what active surfaces were available in prebiotic times? Perhaps the most promising mineral candidates are clays, which are ubiquitous fragments of rock weathering. Clays are loosely layered materials in which stacks of strongly bonded atomic sheets pile up like a ream of paper. What's more, clays in modern soils and ocean sediments are invariably coated with organic molecules that glom on to every available surface.

According to one imaginative scenario, life arose in the hot environment of a deep sea hydrothermal vent, where simple organic molecules, heat energy, and clay surfaces provided just the right conditions. Until more details can be deduced about the specific steps in the process, however, no one can say for sure. Such uncertainty doesn't stop Stanley Miller, a feisty and outspoken critic of such speculation, from dismissing the idea: "The vent hypothesis is a real loser," he told one interviewer. "I don't understand why we even have to discuss it."

In science, as in many other human endeavors, an authoritative rebuff often serves as an effective substitute for facts—though not permanently; eventually a courageous scientist, usually a young one, insists out loud that the emperor has no clothes.

Step 3: Membranes

All cells possess the same kind of outer membrane—a double layer of long, skinny phospholipid molecules that fit side by side, with their water-loving phosphates forming a surface like pencils in a box. If present in sufficient concentrations, these molecules will aggregate into sheetlike structures similar to the membranes that separate the inside from the outside of every cell. Cell membranes are an essential component of every living object. Harold Morowitz, for example, claims that "it is the closure of a . . . bilayer membrane into a vesicle that represents a discrete transition from nonlife to life."

Two variants of the encapsulation process have been proposed. Morowitz is among those who believe that encapsulation occurred rather early in the prebiotic sequence. Such an isolated environment, he argues, provided a safe chemical lab for other reactions to proceed. De Duve, on the other hand, sees the formation of a membrane as the last step—one that followed formation of a self-replicating RNA package. Happily, neither side seems too dogmatic about this issue. Everyone agrees that the membrane had to have formed sometime after simple molecules and before the first cell. In the origin-of-life business, any point of agreement is like a calm harbor in a stormy sea.

Step 4: Self-Replication

Life began when a highly organized collection of molecules began to duplicate itself. No one knows the detailed anatomy of that first self-replicating object, but experts agree that it must have had the ability to obtain matter and energy from its surroundings: the process called metabolism.

METABOLISM: LIFE AS THE ULTIMATE LABORATORY

Perhaps the best laboratory for studying life's origins is life itself. The biochemistry and cell biology of modern cells may well carry the signature of the first cell. The shared characteristics of living cells may point to features present in the earliest living cells. Species, once evolved, cannot readily exchange cellular structures or chemical mechanisms, nor is a specific chemical, structure, or process likely to evolve independently in all organisms. Features shared by all cells thus likely reveal characteristics of the first cell.

The metabolic chemical reactions that convert molecules to energy in even the simplest single-celled organism are vastly complex. But lurking beneath that complexity are shells of more and more basic processes, something like the stacking of Russian babushka dolls. If we could strip away all the outer layers—chemical reactions added on to the simplest chemistry of the first cell—we might glimpse the secrets of that earliest life.

Complexity

Metabolism creates entire cells from carbon dioxide, water, and a few other simple compounds. By examining living organisms, researchers hope to deduce the most primitive metabolic chemical pathways hidden inside. This formidable problem is something akin to describing the workings of the very first IBM computer language by analyzing the software of a modern PC. Each new generation of computer language had to incorporate features of the previous versions in order to maintain continuity. A complex layering of odd conventions and inefficient logical structures were inevitable as the new tried to replace the old in an uninterrupted progression.

Digital computers, the legal system, football rules, and life-forms are examples of complex competitive systems, that tend to evolve in predictably inefficient and strange ways. The constant struggle for survival does not provide the luxury of time to discard layers of inefficiency. In fact, some biologists point to the many apparently silly inefficiencies of living systems—vestigial organs, junk DNA, the odd position of light-sensing cells in the human eye—as perhaps the strongest line of evidence that modern life has evolved gradually from simpler forms.

A Competing Hypothesis

The Miller-Urey experiment provides a spectacular demonstration of how organic molecules might form from water and other molecules. Prior to that study, most organic chemistry was done as far away from water as possible, because water inhibits many of the most interesting and useful reactions. Biologist Harold Morowitz credits Miller and Urey for opening up the field of aqueous organic chemistry. But beyond that, Morowitz claims, these experiments offer little insight. By his estimate, the Miller-Urey experiment has been reproduced in perhaps 5000 variants, each leading to a publication.

"You mix garbage in, you get garbage out, and you have a publication on the origin of life," he laments.

According to Morowitz's somewhat heretical view, detailed in his monograph *The Origins of Cellular Life* and subsequent publications, many origin-of-life researchers are asking the wrong question. It's irrelevant whether amino acids or lipids can be synthesized randomly in small quantities in a Miller-Urey–type environment, he asserts. The important question is by what simple, *self-replicating* chemical pathways can carbon dioxide—ultimately the source of all biological carbon—be incorporated into larger molecules. The random production of a molecule here and there doesn't help. The key is self-replication.

Morowitz believes that he has found such a chemical path in the familiar citric acid cycle, which is deeply buried in the modern-day metabolism of every living thing. The citric acid cycle consists of a few small molecules, including two-carbon acetate, four-carbon malic acid, six-carbon citric acid, plus CO_2. As this cycle proceeds, energy is generated for life's many functions.

What's more, the simple components of this cycle are building blocks for all biological molecules. In a recent key experiment, Morowitz combined one of the cycle's components (a five-carbon molecule called alpha-ketoglutaric acid) and ammonia in a beaker, and amino acids formed spontaneously. No catalyst was required, nor did he resort to fancy electric sparks or ultraviolet light.

Morowitz suggests that encapsulation by a lipid bilayer occurred very early in the process, followed by the self-replicating cycle. Such encapsulation was essential to maintain an effective concentration of organic chemicals. Perhaps this protocell, though unlike modern DNA-based cells, was an entity capable of duplicating itself. RNA chemistry, followed by DNA synthesis, were relatively late-stage developments in the path to life as we know it.

Morowitz's novel idea must be tested and modified, and perhaps eventually discarded as just one of many intriguing but incorrect solutions. Many researchers with vested interests in other theories will criticize or ignore this scenario, while promoting and perhaps testing their own ideas. Thus science progresses—not as an inevitable march from fact to fact, but as a jerky, faltering, joyously unpredictable adventure.

THE LANGUAGE OF LIFE: CAN WE UNRAVEL OUR GENETIC CODE?

It is hard to imagine the impact the discovery of the structure of
DNA had on our perception of how the world works. Reaching
beyond the transformation of genetics, it injected into all of
biology a new faith in reductionism. The most complex processes,
the discovery implied, might be simpler than we had thought.

—Edward O. Wilson, *Naturalist* (1994)

Genetic Information

Without information, there could be no life. To construct a living
thing requires vast amounts of information about molecular building
blocks, body architecture, and the dynamic processes by which organisms develop and reproduce. In a sense, all life-forms represent solutions to the problem of passing this information from one generation
to the next.

At the instant of conception, all the information necessary to create
a unique human being unites in the fertilized egg, an object about a
hundredth of an inch across. As that first cell divides and an embryo
develops, this coded information must be duplicated and read time and
time again. Each individual may eventually pass on a portion of his or
her unique genetic message to subsequent generations. This process of
information transfer has continued unbroken from the first living cell
nearly four billion years ago to the diversity of modern life.

For centuries, scientists and philosophers wondered how the vast
amount of information necessary to produce a living organism could be
stored, duplicated, and interpreted in an object as tiny as a cell. By early
this century, biologists agreed that the genetic message is somehow
carried by chromosomes, elongated cellular structures that duplicate
and divide evenly just prior to cell division. Researchers also concluded
that the information must be coded into some kind of molecular structure—molecules being the only object capable of carrying information
and small enough to fit inside the cell.

Two kinds of molecules, proteins and deoxyribonucleic acid, or
DNA, combine to form chromosomes. Both molecules seemed possible
candidates, but experiments of Oswald T. Avery at the Rockefeller Institute in New York in 1944 provided very strong evidence that DNA
was the informational molecule. A key question remained unanswered:

What is the DNA molecular structure that enables a cell to preserve and duplicate biological information?

Had this book been written a half century ago, that question would have deserved a prominent place. Its answer, formulated for the most part during the dynamic decades of the 1950s and 1960s, is what many researchers consider to be the pivotal discovery in modern science. And as that deep question was answered, new puzzles have taken its place: How is the genetic information of DNA converted into flesh, blood, and bone? And can we learn enough of that language to rewrite the book of life?

THE BOOK OF LIFE

Of all the scientific discoveries of the twentieth century, none has had a more profound effect on our perception of the physical universe and our place in it than the deciphering of the genetic code. Within the span of a half century we have learned that every living organism, from slime mold to sequoias to us, shares the same chemical language of heredity recorded in the form of the molecule DNA. As molecular biologists grow ever more fluent in that language, they are learning to modify an organism's DNA, thus altering life itself. This awesome capability, and the disturbing ethical questions it raises, places us at a historic transition point in an epic scientific journey that has spanned thousands of years of observation and research—a journey that will continue to captivate biologists for centuries to come.

Learning the Alphabet

The molecular structure of the DNA molecule, deduced by James Watson and Francis Crick in 1953, holds the key to its behavior. (*The Double Helix* by James Watson recounts the imaginative science and surprising social context of this historic research effort.) Much tantalizing information was known about DNA by the early 1950s. Chemical analyses had demonstrated that it is composed of equal numbers of three molecular building blocks:

(1) a five-carbon sugar called deoxyribose

(2) a phosphate group (PO_4)

(3) one of four similar nitrogen-bearing molecules called bases: cytosine, guanine, adenine, and thymine

The three building blocks join to one another to form units that are strung together in long chains, often millions of units long. Furthermore, the amount of the base cytosine (C) always equals the amount of guanine (G), and the amount of adenine (A) always equals the amount of thymine (T).

Watson and Crick, relying on X-ray diffraction photographs of DNA obtained with great skill and considerable difficulty by Rosalind Franklin of King's College in London, realized that the molecule must have a spiral or helical shape. Their brilliant deduction, arrived at more by inspired guess and tinkering with models than any step-by-step logic, was that the structure featured a double helix, something like an incredibly long, twisted ladder. Alternating sugar and phosphate molecules form the vertical sides of the ladder, and the bases, attached to the sugars, point from the sides to the center so that pairs of bases, either C-G or A-T, form the ladder rungs.

The power and beauty of the elegant DNA structure lies in its ability to both store and copy vast amounts of information. The four bases, A, C, G, and T, provide a four-letter genetic alphabet capable of conveying any desired amount of information, depending only on the length of the molecule. (In a similar way, the binary "alphabet" of two letters, 0 and 1, can encode any amount of data in modern computers.) The double helix configuration, with complementary bases opposite each other, also facilitates duplication because each side of the structure contains all the molecule's genetic information. By "unzipping" the double strand, two new double helixes can be assembled in its place. Each new double helix is identical to the other and to the original because C always pairs with G, and A always pairs with T. The scientific community, captivated by the perfection of this simple molecular design, embraced the Watson and Crick model immediately upon its publication.

The discovery of DNA's structure marked a historic transition point in the biological sciences. Prior to 1953, studies of genetics at the level of organisms, chromosomes, and molecules were carried out by different groups with very different training and methodologies. The DNA breakthrough unified genetics by providing a conceptually simple chemical basis for storing and duplicating hereditary information.

Genetic Words

All life relies on a cascade of chemical reactions that occurs in every cell. The information encoded in DNA must provide a blueprint for every step of this complex chemical system. Biologists had long suspected that most of the key cellular chemicals are proteins, a diverse array of polymers composed of various sequences of twenty different amino acids. Each distinct sequence of amino acids forms a different protein, and hundreds to thousands of different proteins play various roles in every cell.

Many proteins serve as catalysts or "enzymes"—molecules that facilitate the synthesis or breakdown of other molecules without themselves being changed. Enzymes digest food in the stomach, use digested food to produce heat and energy, and help to assemble all manner of cellular structures. Other proteins serve as vital building blocks, forming hair, muscles, tendons, cartilage, and other structures. The blood protein hemoglobin carries oxygen from the lungs to the tissues (and gives blood its red color). Hormones like the protein insulin help to regulate body processes. In these and other capacities, the shape of the protein molecule is paramount. Different sequences of amino acids fold, twist, and kink in different ways, creating complex three-dimensional shapes that can lock on to other molecules like pieces in a jigsaw puzzle. A single mistake in the amino acid sequence may cause the protein shape to be altered, rendering it unable to perform its catalytic or structural function.

By the 1950s, many of the complex roles of a cell's proteins were beginning to be understood, but a big puzzle remained: How are DNA instructions interpreted and used to synthesize proteins? Some mutations were known to result in the replacement of one amino acid with another in particular proteins. Other clues were revealed in the late 1950s by pioneering genetic research on numerous mutant forms of the simple bacterium *E. coli*. By documenting and comparing different mutants, geneticists surmised that genetic words were likely to consist of three DNA bases in a row, each three-letter word corresponding to one of the twenty amino acids. A gene would then be a DNA segment that codes for one protein by specifying the exact amino acid sequence.

During a decade of intense and often highly competitive research, biologists struggled to deduce details of the process by which a DNA sequence leads to the synthesis of a protein. They discovered that synthesis depends on a third type of molecule, ribonucleic acid, or RNA, which is closely related in structure to DNA. Each RNA is a single-stranded molecule similar to the double-stranded DNA, except that the sugar ribose substitutes for deoxyribose and the base uracil (U) substitutes for thymine (T).

Three forms of RNA participate in protein synthesis. First, messenger RNA copies the base sequence of a DNA segment (a gene) letter by letter. Only one of the two DNA strands is copied. Each messenger RNA is thus a long, single-stranded molecule that carries information about one gene. By using synthetic RNA chains as messenger RNAs, the genetic code was deciphered in the early 1960s Synthetic RNA was added to the protein-synthesizing machinery extracted from *E. coli* cells. The breakthrough came when workers at the National Institutes of Health used a synthetic RNA containing exclusively uracil bases. The unique protein produced was made entirely of the single amino acid phenylalanine. The scientists concluded that UUU (in messenger RNA) or TTT (in DNA) codes for phenylalanine. Additional experiments soon revealed the "spelling" of all the three-base "words" of the genetic code. The complete genetic vocabulary consists of sixty-four different words—all possible three-letter combinations of A, C, G, and T (or U). Three of the "words" (or codons) do not specify any amino acid but mark the end of a gene on the DNA strand. One codon, ATG (or AUG) codes for the amino acid methionine and also marks the beginning of all genes. Most amino acids are specified by more than one codon (e.g., phenylalanine by TTT or TTC). By 1964 the genetic code, a chemical language common to every living thing on Earth, had been cracked.

Two other RNA types—transfer RNA and ribosomal RNA—are important for protein synthesis. Transfer RNA molecules have an amino acid attached to one end of the chain, and three exposed bases in the middle. The three exposed letters match a codon to the appropriate amino acid. Assembly of a protein with thirty amino acids thus requires the participation of thirty transfer RNA molecules. Finally, ribosomal RNA provides the chemical machinery for linking the amino acids together in the order dictated by the messenger RNA (and the DNA gene) and "interpreted" by the transfer RNA. The result is the protein encoded by the gene. (For an amusing and accessible review of this process, see *The Cartoon Guide to Genetics* by Larry Gonick and Mark Wheelis.) The precision of this remarkable translation of a gene's information into a protein underlies the viability of every living thing on Earth.

Genetic Sentences: What Do Genes Do?

Every human being has about 80,000 genes, each a kind of genetic sentence that carries coded instructions for a different kind of protein. One of the greatest challenges in modern biology is to identify each of those genes, along with the structure and function of the associated proteins. This epic ongoing task is still in its infancy, with not more than half of human genes even recognized, and but a handful of those fully described.

Preliminary gene catalogues containing partial sequences of DNA provide hints to the varied functions of proteins. When two genes are observed to have a similar DNA sequence, their associated proteins probably have a similar function as well. A recent survey of about 10,000 human genes suggests that roughly a quarter of the body's known proteins act as enzymes that help to manufacture cell materials, fifteen percent take part in metabolism, and about ten percent form critical cell structures. Other proteins are essential for cellular defense, communication, and reproduction. But the functions of fully a quarter of all known genes, and the proteins for which they code, remain unknown.

Discovering a protein's three-dimensional structure is fiendishly difficult. The "simple" first step is defining the unique sequence of amino acids that is coded by DNA. This protein chain may be anywhere from less than a dozen to tens of thousands of amino acids long (the current record holder is the springy, fibrous muscle protein titin that consists of about 27,000 amino acids). Like a string of words in a foreign language, the amino acid sequence doesn't make much sense until biologists and chemists learn how the chain folds up. Each chain of amino acids kinks, twists, loops, and folds back on itself to form a complex three-dimensional object. Then several protein molecules may have to stick together to make one functional enzyme. While determining the complete structure of a single protein molecule can take a team of researchers years to complete, the effort is nevertheless vital to understanding life. The exact shapes of these large molecules enable them to perform their special functions in a cell.

So far, no one has learned to predict with complete confidence how any given amino acid sequence will fold—a problem viewed by some molecular biologists as *the* key unanswered question in biology. Armed with that knowledge, the shape of any protein—a clue to its function— might be deduced directly from a DNA sequence. A groundbreaking effort by medical researchers George D. Rose and Rajgopol Srinivasan of Johns Hopkins University points to the nature of a possible solution. By compiling data on structures of known proteins, they have developed empirical rules that describe relationships between amino acid sequence and protein shape. They have developed a computer program that, given an amino acid sequence, predicts the folding correctly about 85 percent of the time. As new protein structures are determined, new empirical rules are sure to emerge.

The importance attached to such efforts lies in the close relationship between a protein's folded shape and its cellular function. Direct determination of protein structure by X-ray crystallography is a difficult and tedious process—one that depends on obtaining perfect protein crystals, which are notoriously difficult to grow. A reliable computer program would not only shortcut that step, it would aid immeasurably in creating "rationally designed" drugs that interact with targeted proteins in precise ways.

In the short span of a half century, molecular biologists have deciphered the four-letter genetic alphabet, its complete sixty-four-word vocabulary, and a few sentences. Now we are poised to read the book of life.

READING THE BOOK OF LIFE

The Human Genome Project

Early in the next millennium, less than a decade from today, one of the great scientific challenges will have been met. After many years of effort, scientists will have at last determined a full record of human genetic information—the human genome. The epic Human Genome Project, funded in part with billions of federal dollars and coordinated from a base at the National Institutes of Health in Bethesda, Maryland, addresses a mammoth problem in molecular biology and information technology. Not only must it document all three billion letters of the human genetic message, it must also provide a detailed map for the distribution of genes on each of the twenty-three pairs of human chromosomes. The task is daunting, but, once completed, it will become an indispensable guide for all future genetic research.

Mapping of the genome relies initially on the identification of distinctive short DNA sequences, typically 200 to 500 base pairs long, that are similar if not identical in everyone's genome. This ongoing effort requires the combined skills of molecular biologists, who manipulate and analyze DNA sequences, and information experts, who collate the vast amount of genetic data and search for recurrent patterns. By establishing tens of thousands of these road markers or sequence tags, researchers are creating a reference system to exact physical locations along chromosomes, not unlike the coordinates of a road map. The first 30,000 sequence tags were recorded by mid-1997, and as many as 100,000 markers may eventually be used to identify any location along the human genome.

Sequencing, another side of the Human Genome Project, is more a problem of information gathering and processing than of science. Thanks in large measure to the influx of genome funding, great advances have been made in automated technologies that process strands of DNA and determine the exact spelling of segments of the genome. In fact, the original target date of 2005 for completion has been moved up to 2003, thanks to improved rates and accuracy of sequencing.

The first complete human genome (which, by the way, will be an eclectic collage of sequences from many different people rather than from any one individual), will provide an invaluable genetic baseline. After the sequencing is completed, there remains the mammoth task of determining the function of each human protein. This task will set an agenda for biology throughout the next century and beyond. The effort will be facilitated by knowledge of the genomes of other organisms that can be experimentally manipulated—yeast, worms, flies, and mice. For this reason the Human Genome Project also supports the mapping and sequencing of genomes besides the human genome. In 1996, an international consortium completed sequencing of the entire thirteen-million-base-pair genome of yeast. Comparisons with the human genome will reveal much about the chemistry of development, metabolism, aging, and disease common to all life-forms, as well as the intricate pathways of evolutionary change.

How Does Genetics Relate to Evolution?

The fundamental discoveries of molecular genetics place all of biology in a new, more unified framework, and have led inevitably to unanticipated insights in every aspect of the life sciences. In no field is this revolution more evident than the study of evolution. Genetic material is *the* physical pathway for heredity, while mutations represent the long-hidden mechanism for evolutionary change. Thanks to the universal language of DNA, every cell of every species—bacteria and plants and animals—harbors an unparalleled record of the dynamic process and detailed history of evolution.

At one extreme, the degree of similarity in certain genes common to all living things provides a broad-brush picture of evolutionary branching. The protein cytochrome, for example, plays a key role in the uptake of oxygen in all complex organisms. The amino acid sequence of cytochrome is identical in humans and chimpanzees, indicating a common ancestry. Cytochrome proteins in humans and yeast are similar but not identical; they differ in the identity of 35 of 104 amino acids, suggesting a very ancient common ancestor. By sequencing the cytochrome gene in many different organisms, an evolutionary map spanning hundreds of millions of years of gradual change can be produced. Other genetic changes occur at a much faster pace than those of cytochrome, and thus reveal details of recent human evolutionary patterns. Every human cell contains energy-producing bodies called mitochondria. Each of these cell components contains its own double helix of DNA, which is duplicated as a mitochondrion multiplies. Because mitochondria are passed exclusively from the mother's egg to the child, all individuals with the same maternal lineage have the same mitochondrial DNA.

Genetic analysis of mitochondrial DNA, which mutates relatively rapidly, is the newest tool for anthropologists. Similarities in mitochondrial DNA sequences among the native tribes of Siberia and North America, for example, indicate genetic links between these geographically separated populations. When coupled with other evidence such as detailed linguistic analysis, this research paints a sweeping panorama of human migration and interaction.

Similar DNA tests shed light on long-standing historical mysteries by establishing disputed parentage. In one much publicized case, DNA tests were performed on Anna Anderson, who claimed throughout her long life to be Anastasia Romanov, the only surviving offspring of Czar Nicholas II. The results revealed that she was in fact an impostor, most likely a Polish factory worker who disappeared at about the time of Anna's "discovery."

Genetic research has also revealed a more subtle aspect about the process of mutation and evolution. One extraordinary feature of the molecule-based genetic code is that single random variations in the DNA base sequence have the potential to alter an organism in dramatic ways. In most physical systems, small random variations average out and have no net effect. In genes, however, the smallest of variations—an error in a single genetic letter out of billions—can result in a dramatic mutation that changes the function of a protein and the properties of the organism. While most mutations are harmful to the individual, once in a while a change will confer a competitive advantage. Thus, Darwin's thesis of evolution by natural selection can now be tied to the remarkable behavior of DNA.

Genetic errors, changes in DNA sequence, contribute to evolution when they occur in eggs or sperm cells, but they are also responsible for cancer and thousands of inherited human diseases. Molecular biologists want to learn how to alter disease-related changes in the DNA sequences of particular genes so that the disease may be prevented. As we learn to recognize errors in an individual's genome, a great question facing molecular biologists is not only how to read the book of life, but also how to edit it.

The Individuality of Genomes

We speak of "the" human genome, but in fact each person has a unique genome, except perhaps for identical twins. All humans have the same set of genes arranged in the same way on our chromosomes. But among the three billion base pairs, individual differences exist— perhaps once in a hundred to once in a thousand base pairs. While some of these differences are significant and are responsible for inherited diseases, the vast majority of changes are consistent with a fully normal person. They do, however, account for the fact that each of us is unique; we don't look alike, or think alike, or act alike.

The sweeping drama of molecular genetics lies in its potential to affect everyone's life. Scientists now have the ability to differentiate the unique genetic pattern of almost every human individual, often on the basis of DNA contained in just a few cells. DNA fingerprinting, which makes dramatic headlines in lurid rape and murder trials, has the power to set free the innocent and convict the guilty. Genetic tests expose intimate secrets by providing incontrovertible evidence of contested paternity and family lineage. Detection of certain gene mutations is a reliable indicator of some inherited diseases. In other instances the detection of mutant genes can, for now, indicate only an increased risk of disease.

Our ability to manipulate and analyze tiny amounts of DNA rests on a simple and elegant technique called the polymerase chain reaction, or PCR. PCR relies on DNA's ability to direct its own duplication. Scientists begin by preparing a solution that contains the DNA sample of interest, perhaps the chromosomes from a single cell. The important DNA may be only a very, very tiny proportion of the total. To this solution are added significant amounts of three key ingredients: individual DNA nucleotides (individual rungs of the DNA ladder), an enzyme called polymerase that helps to assemble the DNA strands from the nucleotides, and numerous copies of two short "primer" segments of

DNA that recognize and bond to desired sequences of perhaps twenty bases within a target gene. This solution is heated to about 200°F, just below the boiling point of water—a temperature sufficient to separate double-stranded DNA into free-floating single strands. The mixture is cooled to about 140°, cool enough for the two different primer strands to bind to the respective matching segments on the two exposed single DNA chains in the target gene. Polymerase triggers the growth of two new DNA strands, to be extended from the primers. The result is two new double helixes where before there was one.

The process is repeated over and over again: two strands of the desired gene sequence become four, four become eight—each heating and cooling cycle lasting a few minutes. Within a few hours, millions of copies of the target gene sequence have been produced, while the rest of the cell's genome remains as a single copy. Given this large amount of a single gene, other routine procedures can be used to determine its DNA sequence.

PCR provides the molecular biologist with a powerful tool to identify and understand genetic characteristics. It has found routine use as a rapid screening technique for numerous genetic diseases such as cystic fibrosis and muscular dystrophy, for viral contaminants such as HIV in blood, for tissue compatibility before organ transplants, and for criminal investigations when tiny tissue samples must be processed. PCR amplification of viral and bacterial genes, which mutate rapidly into new variants with slight modifications, now represents an important weapon in tracking disease vectors and controlling outbreaks.

Once a gene has been isolated and amplified by PCR or other genetic engineering techniques, one key to understanding how it works is to document the protein structure for which it codes. In some cases the DNA sequence exactly matches the protein—every triplet of bases matches one amino acid in the protein. But genes are often complexly structured, containing a variety of "editing instructions" that must be undertaken before a gene can become a useful protein template. Many human genes, for example, are broken into several widely separated segments on a chromosome with one or more intervening DNA sequences that are not part of the gene. Cellular mechanisms must remove these intervening sequences and splice the complete gene together before a correct messenger RNA is produced and protein synthesis can begin. Scientists do not understand completely how to read those convoluted DNA editing instructions. Moreover, often the intervening DNA sequences are much longer than the gene sequences themselves. Therefore, it is often very helpful to have a copy of the messenger RNA, rather than the gene itself, to expedite understanding of protein structure.

Fortunately, there is a clever way to make a DNA copy of a messenger RNA strand. Scientists take advantage of the existence of a group of enzymes called reverse transcriptase. These enzymes use a single-stranded messenger RNA molecule as a template to make a DNA copy of the base sequence in the RNA. Each strand of messenger RNA contains the complete instructions for manufacturing one protein. While the original DNA gene may contain intervening sequences, the messenger RNA is an exact and correct genetic message, without all the extra DNA. Molecular biologists extract the messenger RNA from a cell, and then mix it with reverse transcriptase and nucleotides to produce the complementary DNA. A strand of complementary DNA can be amplified by PCR and then sequenced to reveal the protein structure.

Diagnosis

PCR is now a routine genetic tool, but the questions it is helping to answer are far from routine. Physicians desperately want to learn the causes of inherited diseases, of which more than three thousand are now recognized. Each such disease arises from a different gene defect, and each must be investigated individually. In some instances, a single mistake in a gene's nucleotide sequence leads to a defective protein—one that is no longer able to fulfill its intended function. Sufferers of sickle-cell anemia, for example, possess a defective gene for the vital blood protein hemoglobin. One incorrect genetic letter out of 363—T instead of A—results in a protein in which the amino acid valine incorrectly substitutes for glutamic acid at one critical position in hemoglobin. That error alters the shape, and thus the function, of the hemoglobin and leads to the debilitating disease.

In the case of Huntington's disease, the gene defect is related to a variable number of CAG codons, which give long strings of the amino acid glutamine in the corresponding protein. Individuals with healthy genes typically have from 11 to 34 CAG repeats, while individuals with 36 or more CAG repeats are likely to suffer the disease. Furthermore, increased numbers of repeats seem to correlate with earlier onset of the disease.

An extraordinary new genetic technology may result in routine accurate diagnosis of many types of genetic defects. In a process that is now under commercial development, a glass plate, or "chip," is etched with an array of up to 60,000 squares, each no more than about 1/250th of an inch square. A different specific nucleotide sequence about ten bases long is synthesized on each square using computerized techniques like those employed in the computer industry to make microchips. A single chip will contain thousands of variant DNA sequences, both normal and abnormal, of a known gene associated with breast cancer or cystic fibrosis, for example. A patient's gene, amplified by PCR, binds to the appropriate square when put in contact with the chip. The patient's exact variant of the suspect gene, and thus the risk of disease, can be determined.

Someday, as the many variants of each gene are understood, this type of DNA analysis may provide everyone with a useful genetic medical analysis at birth.

CAN BIOTECHNOLOGY REVISE THE BOOK OF LIFE?

Research in molecular genetics has become extraordinarily rich and varied. Broadly overlapping areas of pure and applied research continue to emerge and together employ more scientists than in any other field. Many molecular biologists seek to understand the bewildering array of mechanisms employed by cells to express their genetic vocabulary. Others, by contrast, focus on the manipulation of genetic material—genes and other segments of DNA or RNA—for medical and commercial purposes. As we learn new ways to alter genes, the prospects for genetically engineered products—foods, drugs, pets, and people—seem almost unlimited. Many useful products can come from this research. But like all new technology, genetic engineering also poses problems.

Genetic engineering is the process of consciously altering a coded sequence of DNA or RNA. A remarkable variety of enzymes that cut, paste, rearrange, and copy DNA have been discovered and, together with sophisticated chemical methods, they comprise a powerful tool kit for the molecular biologist. Virtually any desired sequence of nucleotides can now be synthesized and introduced into cells. Furthermore, with some experimental organisms it is possible to remove specific DNA sequences. As we have already seen, the genetic tool kit includes techniques that can be used to compare quickly and accurately DNA from different organisms for screening hereditary diseases, mapping evolutionary pathways, and identifying criminal suspects.

Transplanting Genes

Isolating and identifying genes is one thing, but moving them from one organism to another is quite another. The earliest efforts in genetic engineering, which focused on the large-scale production of proteins for medicine, food production, and basic research, employed single-celled bacteria that reproduce rapidly. In this simplest of genetic engineering protocols, bacterial cells are exposed to strands of DNA with a desirable gene. The DNA may be from any other organism or even a virus. Given enough bacteria, a few cells may incorporate that gene into their own genome and begin to produce the desired protein. Dramatic early success included the development of E. coli strains engineered to contain synthetic genes for key components of human insulin and synthetic growth hormone. Pharmaceutical companies culture these bacteria in large vats to produce the desired proteins in commercial quantities.

On a smaller scale, scientists obtain significant quantities of rare proteins by a similar technique. In particular, pieces of complementary DNA chains produced from messenger RNA can be duplicated in large numbers by PCR techniques, and then introduced into E. coli. The resulting proteins can then be purified for structural studies and other investigations. This procedure has proven especially valuable in studies of defective proteins that intentionally incorporate one amino acid mistake. By understanding the behavior of mutant proteins, biochemists gain insight to the function of enzymes and the mechanisms of genetic disease.

The simple protocol for inducing protein synthesis in a bacterial cell, by inserting the appropriate gene into that cell's genome, also works for many plants. Under the proper growth environment, entire plants can be generated from a single cell—an option not yet possible with animals. In one historic series of experiments, plants were infected with engineered bacteria that contained a gene for proteins that are toxic to insects. The bacteria, which attach to wounded plant tissue, insert some of their DNA into the plant cells, causing a tumor to form. Remarkably, individual plant cells from this tumorous growth developed into a new strain of healthy, insect resistant plants. Continuing experiments and their large-scale applications now focus on increasing crop yields for foods and fiber, extending shelf life of fruits and vegetables, and improving resistance to disease and drought.

Nature has evolved an astounding variety of proteins. As the techniques for modifying organisms with new genes continue to improve, many bioengineers focus their attention on mining the genomes of humans and other organisms for this chemical bounty.

Gene Therapy

In principle, genetic engineering of humans would be the same as in plants and other animals. In practice, however, modifying human genes, even for therapeutic purposes, is a daunting challenge, combining scientific hurdles and moral complexities. One reason for pursuing this challenge is the promise of improved treatment or even cures for presently intractable diseases.

The bright potential of gene therapy was epitomized in the story of Ashanthi DeSilva, who was born in 1986 with a devastating genetic illness, severe combined immunodeficiency. A defect in one critical enzyme, adenosine deaminase or ADA, prevented her immune system from responding to bacterial invaders. Asha's doctors knew that without drastic measures to isolate her from every source of infection, she would probably die in childhood.

With little hope for a normal life, four-year-old Asha became the focus of a pioneering medical experiment in 1990—the first attempt to replace a faulty human gene. Researchers removed some of her blood cells, modified them with a correct version of the gene, and injected the cells back into her body. Six years later Asha was thriving and living a normal life, her immune system was functioning normally, and doctors were cautiously optimistic that her condition has been cured. As the first gene therapy patient, she has also achieved something of a celebrity status. She was named an honorary "research ambassador" for the March of Dimes in 1993, and appeared before Congress in 1994 as "living proof that a miracle has occurred." For a time it appeared that gene therapy would be the "magic bullet" of the 1990s, but as we will see, Asha's improvement was not so easily explained and gene therapy remains a hope, not a promise.

As advances in molecular biology point medical researchers to the precise genetic defects that cause many crippling diseases, it's easy to be optimistic that cures must be close at hand. Cystic fibrosis, muscular dystrophy, sickle cell anemia, hereditary forms of cancer, and numerous other genetic ailments are now understood at the molecular level. However, an important unanswered question in many of these cases is, quite simply, how to repair or bypass a broken gene. Unfortunately, the straightforward genetic approaches that have worked with bacteria and plants are not amenable to us.

Gene therapy through the modification of genetic errors became an exciting, if speculative, area of biomedical research after the first approved human experiments in 1990. Since then, more than 125 clinical trials affecting about 600 patients have begun. The government has contributed about a quarter billion dollars a year to these efforts, most of which are undertaken by private biotech companies.

Gene modification therapies adopt a variety of disease-fighting strategies. About twenty of these trials focus on the introduction of specific normal genes. Most of the other trials are efforts to make cancer cells more vulnerable to attack by inserting a gene for a protein that turns on the immune system. Still other gene therapy research attempts to thwart the AIDS virus by introducing genes for the proteins that coat HIV; by priming the patient's immune system to attack HIV proteins, an effective AIDS vaccine might result. In all of this research, scientific issues are clouded by the emotional response to dying children and the large financial return if a successful cure is found.

There's nothing particularly mysterious about the preferred approach to gene therapy, at least in principle. While treatment of each disease differs in detail, the basic strategy is the same in every effort:

Insert a normal gene into genetically defective cells. Several creative strategies have been attempted. In some instances, defective cells, for example from the blood or bone marrow, can be removed, modified by inserting the correct segment of DNA or complementary DNA, allowed to multiply, and reintroduced into the organism. The most difficult part of this process is insertion of the corrected gene, which must be accomplished by a "vector" that acts as a molecular delivery system. The most common vectors are built from retroviruses—viruses with RNA genomes. The genomes are engineered to include messenger RNA corresponding to the normal version of the cell's defective gene. Retroviruses survive by converting their RNA genomes into DNA (with reverse transcriptase) and inserting the DNA into the target cell's genome, which then are tricked into producing the viral proteins.

With only a few years of research under their belts, no one in gene replacement therapy research can predict where these experiments might lead. Biologists at AntiCancer Inc. in San Diego are studying the possibility of gene therapy for hair loss, though a substantial roadblock is that no one has discovered exactly which gene or genes causes male baldness. But who knows, perhaps someday you'll be able to buy a genetic shampoo. On a more sobering note, some geneticists talk about engineering a forty-seventh human chromosome—a designer genetic cure-all that could introduce an array of new "desirable" genes into a patient.

Despite all the hype surrounding gene therapy, not one of the many clinical trials has yielded definitive positive results, and many, such as proposed cures for cystic fibrosis, have been out-and-out failures. In a sobering assessment, the National Institutes of Health issued a report in late 1995 that concluded "clinical efficacy has not been definitively demonstrated at this time in any gene therapy protocol . . . despite anecdotal claims of successful therapy." Even the much-publicized case of Ashanthi DeSilva is ambiguous, for she has been receiving regular injections of synthetic ADA in addition to gene therapy.

One major difficulty of all gene-insertion techniques is that with only a few exceptions in experimental organisms, DNA introduced into a cell's chromosomes is inserted essentially at random. Given enough bacteria or plant cells, some of the engineered genes will likely end up in a fortuitous position where they can be expressed. Other genes will be ignored or attacked by the cell's complex mechanisms that repair errors in the DNA sequence, and still other DNA inserts will interrupt a critical cellular gene, and thus kill the host cell. Nevertheless, the most successful cells can often be separated from the others and used in future experiments. But with animals, the only hope is to modify enough cells so that some will produce the missing enzyme.

Gene therapy on humans affects specific groups of cells in the body; it does not produce genetic alterations that can be inherited. Ultimately, such alterations may be desirable to eliminate a hereditary disease. They are already being used in efforts to create improved breeding stocks in agriculturally important animals. In order to create such permanently altered or "transgenic" organisms, sex cells must be modified to incorporate the changes that are to be passed on to subsequent generations. Such modifications, which have become routine in only a few experimental animals, notably fruit flies and mice, involve injecting the modified genetic material—a transgene—into many fertilized eggs. In a few eggs the new DNA may be incorporated into the host's genome, though little control is exerted on exactly where. With luck, mice with a predisposition to specific types of cancer may be produced. Furthermore, by locating the site of the transgene in many different strains, it may be possible to document aspects of gene expression, regulatory proteins, and tumor formation.

The introduction of transgenic organisms raises serious ethical questions. Most people consider such efforts acceptable if applied toward animals used in cancer studies and other pressing biomedical research. But the possibility of transgenic humans is far more troubling.

Can We Bring Back the Dinosaurs?

The memorable premise of *Jurassic Park* begins with dinosaurs at the smallest scale—the scale of a single cell. Ancient bloodsucking insects, so the story goes, preserve dinosaur blood cells, each with a set of dinosaur DNA. All scientists need to do is collect the preserved insects from amber mines, extract dinosaur cells from their gut, and use the DNA to grow the mighty beasts.

Is this idea possible? Each step in the Jurassic Park scenario does, indeed, have a tenuous basis in fact. Fossil insects are often beautifully preserved in amber, though they do not contain pristine dinosaur blood cells. A few scientists have reported the extraction of ancient DNA from samples of well-preserved fossil dinosaur bone, but such DNA is invariably broken into small fragments and may be from some organism other than the dinosaur. In the unlikely event that biochemists obtain a complete set of dinosaur chromosomes, it might be possible to insert the chromosomes into a modern egg. But we do not know how to trigger or sustain its development and growth. Consequently, the engineering of an authentic living dinosaur is not a realistic goal for the foreseeable future.

The most basic problem in bringing an ancient life form back from extinction is obtaining its genome—a complete set of its chromosomes. For DNA to work properly, every base pair in every gene sequence must be correct. But the reality is that all DNA strands tend to degrade over time. Background radiation, chemical reactions, and the effects of heat disturb some sequences. The natural decay of radioactive carbon-14 into nitrogen-14, a process that takes place constantly in every living thing, causes other disruptions. The living body has many repair mechanisms that continually reverse this damage, but as soon as an organism dies, its DNA is subject to steady, unchecked decay unless it is kept cool and dry. Ancient bone or blood cells may contain some DNA, but these molecules will be present typically as short, broken segments perhaps only a few dozen bases long. We have no way of merging these bits into a complete genome of several billion bases. Dinosaur DNA may be forever beyond our grasp.

A more accessible, though still distant goal might be to obtain an intact genome from a fossil mastodon or mammoth from material frozen in the Siberian wilderness. Several ancient carcasses, perhaps five to ten thousand years old, have been recovered. Not only do these remains provide trillions of cells to work with, but those cells have also been kept cold and they are relatively young. Essentially intact chromosomes thus might be recovered or reconstructed from these extinct beasts. Furthermore, libraries of genetic material can be stored for any organism now threatened with extinction. Whales, condors, tigers, and many endangered rain forest species could be preserved in the form of a few cells, each with a complete copy of their unique genomes.

Of course, a set of viable chromosomes is not necessarily a viable organism. We have much to learn about the conditions that foster growth of a fertilized egg. Even if we knew how to insert exotic genetic material into the egg of a different species, we have no way of ensuring that an embryo will develop properly.

In the case of genetic engineering, creative imaginations far outpace our technical abilities. But many of the present limitations are technical details, ones that many biologists expect will be worked out in coming generations. Then society will be faced with ethical questions of awesome implications. Someday, when we have learned the language of genes and their regulation, humans may gain the ability to design new organisms. Given the youthfulness of molecular genetics and the extraordinary pace of discovery, it seems inevitable that someday we will be able to modify human traits and abilities, and produce clones of any imaginable organism with desirable characteristics of our choosing.

Then the central unanswered question of genetics will not be "What is the language of life?" but, rather, "What limits must we place on its use?" In recognition of this knotty problem, fully five percent of funding for the Human Genome Project, more than 100 million dollars, has been earmarked for study of the ethical implications of genetic research—the largest ethics program in human history. Let us hope it will be enough.

Evolution: How Did Life on Earth Become So Varied?

Tracking Down an Ancient Clue

For more than a decade during the 1970s and 1980s paleontologist
Peter Sheldon combed the hills of south-central Wales for the most
prized of invertebrate fossils—the elegant, ancient trilobite. Well-preserved trilobites are relatively scarce, and finding just one or two complete ones, with their bumpy head, segmented body, and grooved tail,
constitutes a good day for most collectors. But Sheldon accumulated
trilobites on an epic scale. Collecting in cliff faces and gullies near
lovely towns with euphonious names like Rhayader, Abergwesyn, and
Llandrindod Wells, he unearthed 14,888 of the precious specimens.
Summer and winter, rain or shine, Sheldon excavated layer after layer
of rocks that were laid down as ocean sediments some 450 million years
ago.

Trilobite collecting is an old and honorable pastime for professionals
and amateurs alike; in some localities these valuable relics are even
mined for sale to rock shops and museum collections. Many collectors
have visited the classic Welsh localities and come home with an ancient
souvenir. Sheldon's hoard was different: Not only was it huge, each
specimen was meticulously documented as to its exact stratigraphic
level from within a thick sequence of rocks representing three million
years of Earth history. Having amassed a suite of fossils that was unprecedented in its continuity over such a time span, he proceeded to
count, specimen by specimen, the number of segments that made up
each trilobite's body. For a full year of labor he sat at a microscope and
counted trilobite segments—tens of thousands of them—to see how
the number changed with time.

You might think that such esoteric research as counting trilobite
segments would end up as an obscure monograph destined for forgotten dusty corners of a few specialized libraries. But when it finally
appeared in December 1987, Sheldon's study won a prized slot in *Na-*

ture, one of the premier venues for top-notch science. His Herculean effort commanded attention because it shed light on one of the biggest questions in science: How has life on Earth changed with time?

THE FACT OF EVOLUTION

Charles Darwin proposed that evolution occurs because of the constant struggle for survival. Many more individuals of most species are born than can possibly survive. In the brutal competition for limited resources, individuals with advantageous traits are more favored to survive long enough to pass those traits on to offspring. This view stresses the important fact that the members of a single species are only rarely identical. You can confirm this fact easily the next time you are in a crowd. Over spans of many generations, this natural selection of desirable characteristics over less desirable traits causes organisms to evolve. Every living thing, including Darwin's Victorian readers, were descended from more primitive organisms.

When these radical ideas were first published in 1859, they created a firestorm of protest and indignation. In a famous public debate sponsored by the British Association for the Advancement of Science in 1860, Bishop Samuel Wilberforce mocked his Darwinian opponent, Thomas Henry Huxley, by asking whether it was on his grandfather's or his grandmother's side that the ape ancestry was to be found. Brushing aside this barb, Huxley directed a scathing retort: "A man has no reason to be ashamed of having an ape for his grandfather. If there were an ancestor whom I should feel shame in recalling it would be rather a man—a man of restless and versatile intellect—who, not content with an equivocal success in his own sphere of activity, plunges into scientific questions with which he has no real acquaintance, only to obscure them by aimless rhetoric, and distract the attention of his hearers from the real point at issue by eloquent digressions and skilled appeals to religious prejudice." Huxley explained to his audience that Darwin's remarkable thesis is a *scientific* theory, one that makes specific, testable predictions about past, present, and future life on Earth.

Indeed, Darwin's bold theory has by now been supported by a diverse and still-growing body of observations. Fossils of different ages display logical transitions from less complex (in older fossils) to more complex (in younger fossils): from single cells to multicelled organisms; from soft-bodied animals to those with shells; from exclusively ocean-dwelling life to life on land, in the air, and every other imaginable habitat. More recent fossils retain vestigial characteristics of earlier forms: Humans possess tailbones, penguins have wings, and fossils of ancient whales display tiny hind legs. Even today, biologists are able to observe and document the ceaseless process of natural selection in the subtle changes of living things: In a matter of decades moth populations in Britain have changed color from light to dark to light again to match the variations in soot-stained tree trunks where they rest. Similarly, the gracefully curved beaks of some birds in Hawaii have become shorter in response to changes in their food supply, while new antibiotic-resistant strains of bacteria arise to confound the medical community. In all of these ways, life-forms have evolved, and continue to evolve, to meet the challenges of their changing environments. The implication is that life on Earth has a common ancestry.

Equally compelling evidence for the common ancestry of all life on Earth comes from studies of modern organisms at the cellular, molecular, and genetic levels. All living things are made from one or more cells, all of which display similarities in their external membranes and internal structures, and all of which employ similar chemical mechanisms to process energy and reproduce. And, most remarkable of all, every known organism relies on the exact same genetic code—a chemical language that carries the blueprint for constructing essential proteins in mushrooms, moss, and man. This is why the simplest bacterium readily produces growth hormones for cows, or insulin for humans, when instructed by bovine or human genes.

Even more astonishing to scientists and interested lay people alike is the fact that many essential genes, and the proteins they construct, are very similar in different organisms. Genes from human cells can, for example, replace mutant genes in yeast cells. Genes that are important for laying down the patterns of a fly's body have cousins in mammals that are similar in both structure and function. Thus DNA structure points to the common ancestry of living organisms. Moreover, our current understanding of the continual changes that occur in DNA sequences provides a satisfying explanation for the differences among individuals of a species that Darwin postulated.

Evolution is a fact. Important questions about the mechanisms and rate of evolution, however, as well as intriguing puzzles about evolution's dark companion, extinction, are still matters of much debate.

The Diversity of Life on Earth

Humans have been describing life on Earth for thousands of years, and during the past two centuries have catalogued approximately one and a half million different living plants, animals, and other organisms according to the logical Linnaean system. Yet, by most estimates many millions of life-forms remain undescribed. Most of these creatures are insects or microscopic organisms, though every year one or two new mammals or birds are discovered in isolated valleys or remote jungle tracts. Other strange forms have been discovered recently in places where no one expected to find life: giant tube worms clustered near volcanic vents on the ocean bottom, organisms that thrive miles underground in ancient sediments, and bacteria that by some accounts come to life after lying dormant for millions of years in salt deposits. Before scientists can fully understand evolution, they must document the incredible diversity of life-forms that now inhabit Earth.

Even if scientists could record every living thing, the richness of life on Earth today represents only a tiny fraction of all the forms that have lived. The 3.8-billion-year fossil record reveals a boggling menagerie of extinct creatures: spiny trilobites, coiled ammonites, armored fish, giant dragonflies, imposing dinosaurs, and primitive humans. Oceans once thrived with countless species of brachiopods, bryozoa, blastoids, crinoids, graptolites, and numerous other creatures all but unknown today.

A continuous evolutionary tree ties these diverse forms to a single ancient living cell. But how did the diversity arise, and how is it changing today? To answer these questions requires meticulous examination of Earth's record of past and present life.

THE MECHANISM OF
EVOLUTION

There was a time before Darwin when many scientists, quite sensibly, saw creatures evolving with clear purpose and resolve. The extended neck of the giraffe, the efficient flippers of the seal, and the keen eyes of the eagle were rationalized as end stages of directed evolution.

Darwin and Alfred Russel Wallace, who came to the same conclusion independently at the same time, disagreed. Traits, they argued, vary from generation to generation by the undirected process of random variation and natural selection. If an individual possesses some advantage that increases its chance to survive long enough to reproduce, then that advantage stands a better chance of being passed on. Any advantageous trait that we see in animals or plants today must have arisen through a series of intermediate steps, each of which had to confer some advantage to the individual.

The Evolution of Eyes

Certain key traits seem universally likely to confer advantage. Light-sensing ability, for example, proves useful to all animals except perhaps those in permanently darkened caves or the deepest ocean bottom. Even bacteria can sense light. The fossil record and modern organisms reveal that eyes have evolved independently over and over again during the history of life—as many as sixty different times by some estimates. Critics of Darwin challenged his theory by asking what conceivable intermediate steps could lead from no eyes to the complex organs of insects, octopi, or vertebrates. "What is the use of half an eye?" creationists ask even today.

For more than a century after Darwin's thesis, biologists struggled with the question of how fast and by what mechanisms the transition from sightless to eyed organisms occurred. Ongoing research with computer models is now providing some surprising answers. In one elegant research program, biologists Dan Nilsson and Susanne Pelger of Lund

University in Sweden model eye evolution by making a few simple assumptions:

(1) All eyes begin with one or more light-sensitive cells (something like the photovoltaic devices described in Chapter 5), sandwiched between a transparent protective layer and an opaque backing. This assumption formalizes what biologists have long known; many primitive eyeless life-forms, even one-celled organisms, are sensitive to light. The pigments that sense light are similar in all organisms and the very same pigment is used by all animals, regardless of the varying eye structures. Moreover, a gene called *Pax-6* is essential for initiating eye formation in animals as different as flies and mammals.

(2) Natural variations of about one percent in the size, shape, and distribution of these cells creates a random distribution of curvature, thickness, and ability to bend light. Similar degrees of natural variation are present in all organisms.

(3) From generation to generation, variations in traits that improve light-gathering ability or visual acuity are preserved. The effectiveness of any given eye configuration can be calculated easily from elementary optics, so an objective measure of the relative advantage from one random step to the next is possible.

Nilsson and Pelger's simple computer program modeled the evolution of an incipient eye, step by step, through a random selection process. The results are remarkable. Invariably, within a few hundred generations a well-defined eye cavity, which increases the amount of light detected and decreases the effects of glare, has formed. Shortly thereafter, a lens of thicker cells spontaneously evolves. Remember, each step involves at most a random one percent change in configuration, but such small improvements gradually accumulate. Within about two thousand generations (a geological blink of an eye for most species), a fully developed eye has appeared from a small patch of light-sensitive cells. Their conclusions: "Even with a consistently pessimistic approach, the time required becomes amazing short," from a few hundred years in flies to a few tens of thousands of years in mammals.

Eventually, such quantitative models may be applicable to all sorts of evolutionary transitions. Just as high-tech computer graphics can morph one animal into another (a visual technique used stunningly in *Terminator 2* and other sci-fi and fantasy motion pictures), so, too, will these models reveal "missing links" between diverse forms.

Fast or Slow Evolution?

In an effort to reduce a difficult and complex question to its essence, scientists at times tend to oversimplify nature. Such has been the case more than once in studies of the history of the Earth. Two centuries ago a debate raged between catastrophists and uniformitarianists, who asked, "What is the dominant mechanism of geological change?" The former claimed that epic events such as the biblical deluge shape the planet, while the latter argued for gradual and steady change over un-imaginable eons. Similarly, neptunists and plutonists reduced the complex issue of rock genesis to a question of the action of water versus the action of heat. In each case, the models that seem closest to the mark combine these extremes.

A similarly polarized debate has enlivened recent studies of evolution. Darwin's original thesis, and the point of view supported by evolutionary "gradualists," is that change occurs slowly and in small increments. Such changes are all but invisible over the short time scan of modern observations, and, it is argued, they are usually obscured by innumerable gaps in the imperfect fossil record. Gradualism is a comforting, steady state model, repeated over and over again in generations of textbooks. By the early twentieth century the question about evolution rate had been answered in favor of gradualism to most biologists' satisfaction.

Sometimes a closed question must be reopened as new evidence, or new arguments based on old evidence come to light. In 1972 paleontologists Stephen Jay Gould and Niles Eldredge challenged conventional wisdom with a dramatic variation on natural selection. This opposing viewpoint, called punctuated equilibrium, posits that species change one to another in relatively sudden bursts, without a leisurely transition period. These episodes of rapid speciation are separated by relatively long static spans during which a species may change hardly at all.

The punctuated equilibrium hypothesis attempts to explain a curious feature of the fossil record—one that has been familiar to all paleontologists for more than a century, but has usually been ignored. Many distinctive fossil species appear to remain unchanged for millions of years—a condition of stasis at odds with Darwin's model of continuous change. Numerous distinctive marker or "index" fossils, for example, are abundant throughout the world's rocks, and prove useful in dating sediments. Intermediate fossil forms predicted by Darwin's gradualism, however, are for the most part lacking. In most localities a given species of clam or coral persists essentially unchanged throughout a thick formation of rock, only to be replaced suddenly by a new and different species. Furthermore, similar discontinuous sequences of distinctive organisms are commonly found in several widely separated localities.

The evolution of North American horses, which was once presented as a classic textbook and museum example of gradual evolution, is now providing equally compelling evidence for punctuated equilibrium. A convincing fifty-million-year sequence of modern horse ancestors, each slightly larger, with more complex teeth, a longer face, and more prominent central toe, seemed to prove Darwin's contention. But close examination of those fossil deposits now reveals a somewhat different story, one that is wonderfully displayed in the recently rearranged exhibit of horse evolution at New York's American Museum of Natural History. Horses evolved in discrete steps, each of which persisted almost unchanged for millions of years and was eventually replaced by a distinctive newer model. The four-toed *Eohippus* preceded the three-toed *Miohippus*, for example, but North American fossil evidence suggests a jerky transition of distinct species between the two genera. If evolution has been a continuous, gradational process, one might expect that almost every fossil specimen would be slightly different from every other.

If it seems difficult to imagine how major changes could occur rapidly, consider this: An alteration of a single gene in flies is enough to turn a normal fly with a single pair of wings into one that has two pairs of wings.

The question of evolution rate must now be turned around: Does evolution ever proceed gradually, or does it always occur in short bursts? Detailed field studies of continuous sequences of fossiliferous rocks provide the best potential tests of gradualism versus punctuated equilibrium. Occasionally, a sequence of fossil-rich strata permits a comprehensive look at one type of organism over long time spans.

Peter Sheldon's trilobite localities in central Wales, for example, offer a detailed glimpse into three million years of one marine environment. In that study, each of eight different trilobite species was observed to undergo a gradual change in the number of segments—typically an increase of one or two segments over the whole time interval. No significant discontinuities were observed, leading Sheldon to conclude that environmental conditions were quite stable during that period.

Similar exhaustive studies are required for many different kinds of organisms from many different time intervals. Most researchers expect to find that both modes of speciation are correct: Slow, continuous change may be the norm during periods of environmental stability, while rapid speciation occurs during periods of environmental stress. But a lot more studies like that of Peter Sheldon are needed before we can say for sure.

How Did Humans Evolve?

Few scientific questions have been more avidly watched, widely reported, or hotly debated than the evolution of our own species. Theories of our ancestors' lifestyles and behavior make great press copy, while reports of ever older protohumans are trumpeted like athletic world records.

Fossil evidence of ancient hominids is so rare and fragmentary that any model of human origins demands the most meticulous examination of the spotty remains, accompanied by a healthy dose of skepticism, creativity, and intuition. The record as we now know it suggests that the first bipedal humanlike creature appeared in Africa more than four million years ago, while the first member of our genus, *Homo*, dates from about 2.5 million years. Various members of the genus *Australopithecus*, with a brain about the size of the modern chimpanzee, appear to lie intermediate between the great apes and us on the evolutionary tree. The quest for more hominid bones, and the publicity that such finds invariably attracts, continues to drive competing teams of paleontologists in northern and eastern Africa.

One of the most intriguing suggestions to arise from questions of human origins and evolution rates comes from Yale paleontologist Elizabeth Vrba, who has studied transitions in dozens of genera of African antelopes. These wonderfully varied mammals underwent a rapid period of diversification at the same time the first of the *Homo* species appeared, about 2.5 million years ago. These events coincided with the onset of a great ice age—a period of relatively sudden global cooling and environmental stress.

One consequence of the rapid cooling, Vrba suggests, is that natural selection would have tended to favor individuals who used heat most efficiently, for example by keeping body size small. In the case of humans and other primates, two distinctive stages of growth are observed: During juvenile growth the body grows slowly relative to the head, while the reverse is true for adolescent growth, when the body grows rapidly relative to the head. At the beginning of the ice age, natural selection may have favored individual protohumans who had a relatively longer juvenile growth stage—a heat-conserving trait that, coincidentally, led to a much larger brain size. If humans underwent this process, as Vrba suggests, then the dramatic increase in brain size that characterizes the genus *Homo* may have been an incidental adaptation to climate, not a specific adaptation for success in tool making or reasoning. Vrba's idea stresses another aspect of evolutionary change: Diversity can be generated by differential rates of development as well as by altered traits.

EXTINCTIONS

Charles Darwin's theory of evolution by natural selection, together with the genetic explanation for the continual generation of variation, is the most powerful explanation for the physical process by which new species—including ourselves—come to be. Some people mistakenly view that inexorable process as an inevitable progression from simpler life forms to us. Nature has arrived at *Homo sapiens* and, as far as many humans are concerned, the process can be put on hold for a while. But that's not the way natural selection works.

Hand in hand with the evolution of any new species must go the extinction of old ones. The vast majority of fossil species, including dinosaurs, trilobites, and countless other bizarre creatures, disappeared millions of years ago. More than 99 percent (by some estimates more than 99.9 percent) of all species that ever lived have vanished.

Occasionally, as when the dinosaurs died out 65 million years ago, a global catastrophe may be responsible for a mass extinction event. In such an episode, many of Earth's life-forms may be killed in a geological instant—anywhere from hours to a hundred thousand years. The scale of the worst of these events, which took place at the close of the Permian period some 245 million years ago, is almost beyond imagining. In a brief geological interval less than a million years, as many as 96 percent of all living species are estimated to have vanished. At least five major extinction episodes and perhaps a dozen more lesser ones, each documented by careful scrutiny of the fossil record, reveal that life on Earth is both vulnerable and resilient.

Death by Asteroid

For most of the twentieth century uniformitarianism, the belief that "the present is the key to the past" was as close to conventional wisdom as anything in geology. According to this paradigm, the Earth's surface features, including life, have been shaped almost exclusively by slow change over immense spans of time. Even natural disasters like volcanoes and earthquakes gradually lead to large changes in thousands of modest increments.

In 1980 a team of Berkeley scientists challenged this accepted view by proposing that a single extraterrestrial force—the impact of a comet or an asteroid several miles in diameter—killed the dinosaurs, along with most other species, at the Cretaceous-Tertiary (K-T) boundary, about 65 million years ago. Their initial evidence was a dramatic enrichment of the rare element iridium in a narrow band of clay deposited right at the critical transition. That concentration, they argued, was a sure signature for extraterrestrial debris, because iridium is thousands of times more concentrated in some meteorites than it is in most crustal rocks.

The Berkeley result initiated a firestorm of indignant protest. Invoking destruction from outer space was ludicrous—more like bad science fiction than good science, opponents argued. But the thesis was soon bolstered by additional chemical and physical tests, including dozens of additional iridium measurements from around the world, distinctive patterns in the isotopes of the element osmium, shocked minerals characteristic of impact events, and a global layer of soot, presumably from massive wildfires. No known terrestrial event could cause such odd features.

The real clincher has come with the identification of an immense, though largely eroded, impact feature more than one hundred miles in diameter and centered near Chicxulub on the Yucatán Peninsula of Mexico. Recent studies date this crater at about 65 million years. The death-by-asteroid concept is now almost universally accepted by the scientific community (it has even become mainstream textbook fare). Given this new perspective, it is not surprising that subsequent studies have revealed other iridium anomalies associated with other extinction events at ages of about 38, 165, and 365 million years. This episode illustrates how quickly scientists are willing to abandon cherished preconceptions in the light of convincing data.

Other Causes

It is hard to imagine an event more catastrophic than a giant comet or an asteroid wiping out most life-forms on Earth. Nevertheless, the most devastating mass extinction of all, the 245-million-year-old Permian-Triassic event, may have been the result of more gradual terrestrial causes. Although the late Permian and early Triassic fossil record is extremely sparse, there is no clear evidence for a large impact. Instead, this episode seems to have spanned about a million years of environmental stress. By the end of the Permian period, all of the Earth's major continents had merged into one supercontinent, causing drastic alterations in global climate, as well as significant changes in the transfer of vital nutrients from land to ocean. At the same time, global sea levels dropped by about three hundred feet, perhaps as a result of the buildup of polar ice or an increase in the volume of the ocean basins. Other researchers point to the possibility of a sudden poisoning of the ocean by a rapid increase in carbon dioxide, now manifest as massive limestone deposits dating from that era.

Added to these dramatic changes was a million-year period of intense volcanism in Siberia—an episode that may have spewed out a cubic mile of lava every year, while it polluted the atmosphere and darkened the skies. Such a scenario may help to explain a subtle and disturbing feature of carbon atoms contained in fossils near the Permian-Triassic boundary. Almost all atoms of carbon (the element with six protons) have either six or seven neutrons in their nucleus, yielding the isotopes carbon-12 and carbon-13 respectively. Photosynthetic plants tend to concentrate carbon-13 preferentially, so the ratio of carbon-13 to carbon-12 in fossil animals reflects the amount of photosynthesis that was taking place: A high ratio means photosynthesis was proceeding efficiently. Curiously, measurements of 245-million-year-old fossils reveal that the ratio may have dropped precipitously during the million years before the boundary. It appears that microscopic plankton, a vital starting point of the ocean's food chain and the organisms that account for most of the Earth's photosynthesis, may have largely died off. A messy and lethal combination of climate change, sea level change, increased carbon dioxide, and intense volcanism thus may have killed the last of the trilobites along with numerous other creatures.

These results have inspired a closer look at the 65-million-year-old K-T boundary. Emerging evidence suggests that many species of plants and animals, including numerous trees and shrubs and about half of all dinosaurs, disappeared during the hundred thousand years before the K-T boundary. Surprisingly, it now appears that a period of climate change and intense volcanism (this time in the Deccan region of India) preceded that mass extinction too. Perhaps a global environment, weakened by millennia of stress, was more susceptible to the devastation of an asteroid. According to this version of the theory, an extraterrestrial impact provided the dramatic end to the dinosaur story, but it was by no means the whole story.

As if all those disasters weren't enough, astronomers imagine at least one other plausible cause for past—or future—mass extinctions. Any star significantly larger than the sun will end its life in a supernova explosion that unleashes frighteningly intense radiation for a short time period. These events are infrequent, but the universe has lots of time. If any of the multitude of large stars within a few tens of light-years of Earth were to undergo such a climactic end, the effects on our planet's biosphere might be severe. While there is no evidence that such an extinction event has ever occurred during the history of life on Earth, the possibility remains.

Whatever the cause of mass extinctions, much research remains to be done on the phenomena surrounding these events. One of the most basic unanswered questions concerns the characteristics of species that survived each event compared to those that vanished forever—the winners and the losers. Detailed studies of this kind will reveal the nature and extent of environmental stresses that followed the impact. For example, studies of marine animals before and after the K-T extinction suggest that deep-sea creatures fared somewhat better than those in shallow coral communities, but a clear picture of the battered ecosystems that survived has not yet emerged. It is possible, as David Raup has suggested, that mass extinctions are somewhat indiscriminate in their killing. Some species may experience "death by bad luck" rather than by the more directed process of natural selection.

Whatever the mechanism of mass extinction, the fossil record demonstrates abundantly that each episode of death was followed by a period when life blossomed anew—an explosion of new life-forms filled the ecological niches then emptied of the old. The rapid evolution of mammals between 60 and 65 million years ago is just one example of this recurrent phenomenon. Detailed studies of life immediately *after* an extinction event, therefore, might reveal much about the rate and pathways of evolution.

Are Mass Extinctions Periodic?

Most mass extinctions appear at first glance to be random instances of devastation by rogue asteroids or comets or massive terrestrial alterations. Such chance events seem inherently unpredictable—the occasional really bad day. Growing evidence, however, suggests that a subtle periodic pattern may underlie these global disasters.

The search for regularly repeating extinction events is an exhaustive and controversial one, demanding meticulous examination of the last 600 million years of the fossil record. During that time, life evolved from the first simple sea creatures with hard parts (and therefore easily fossilized skeletons) to the diversity of animals and plants today. The scientists who describe mass extinctions must comb every facet of that fossil record—the accumulated research of thousands of scientists, collected in the monographs and periodicals summarizing more than two centuries of research. Dinosaurs, worm burrows, tree leaves, starfish, shark teeth, corals, insects, ferns, shells, frogs, wood—the rich history of life is preserved in countless millions of fossil specimens, and described in the vast literature of paleontology. Libraries have become the favored field areas for those who would find mass extinctions.

David Raup and Jack Sepkoski, University of Chicago paleontologists who have devoted more than two decades to the search, initially suggested, in 1983, the presence of extinction patterns. Their work sent a shock wave that still resonates through the scientific community. At the time their "death star" hypothesis was formulated, neither Raup nor Sepkoski had spent much time in the classic academic pursuits of specimen collecting and describing new species.

David Raup is an odd sort of paleontologist; he has never spent a season doing fieldwork, nor has he ever described a new fossil species. He is more at home developing computer programs and formulating big ideas. Nevertheless, his pioneering work in applying mathematical modeling and statistical analysis to the study of fossils has revolutionized paleontology. One of his most elegant creations is a computer program that can be used to model and draw the full range of beautiful spiral shell shapes exhibited by gastropods and other coiled animals. After obtaining degrees from the University of Chicago and Harvard, he became part of the dynamic young faculty of the University of Rochester, where he was joined by his younger colleague, John Sepkoski, Jr. Their collaborative efforts shifted to the University of Chicago in the late 1970s.

Jack Sepkoski, a student of Stephen Jay Gould's at Harvard University, was an avid fossil sleuth. But, like Raup, much of Jack's professional studies have taken place far from exotic field areas (though he still manages to do some fieldwork on Cambrian trilobites). For years he buried himself in libraries, plugging away at his pet project—an exhaustive catalogue of all the thousands of families (groups of closely related genera, which are in turn groups of closely related species) in the marine fossil record. Jack's *Compendium of Fossil Marine Families*, which records each family's name and age span, first appeared in 1982, and it was computerized in 1983. That's when the fun began. In the words of David Raup, "For me . . . , [it] was like a new toy."

Raup and Sepkoski wrote simple computer programs to analyze large-scale trends in the history of life on Earth. Perhaps the simplest question they could ask was how the number of extant families varied with time. Was diversity steadily increasing? Were the so-called mass extinctions statistically significant? Were they clustered in some obvious way? In an effort to remove "noise" from the mass of data, they culled out any family that had survived to the present day, as well as those for which conditions of preservation or rarity cast doubt on their full life span. What emerged from their analysis of the remaining 567 families was a strikingly regular pattern of extinctions every 26 million years. If mass extinctions were purely random events, then their spacing would be random; two or three extinctions close together might be followed by a hundred million years of stability. Instead, the fossil record revealed eight distinct episodes of death spanning the past 245 million years. Statistical tests indicated that the chance of this periodic pattern being a random occurrence was less than one in a hundred.

This exciting and controversial work, engagingly described in Raup's personal narrative *The Nemesis Affair*, sparked controversy and widespread media attention. Paleontologists attacked the conclusions on several grounds: The geological time scale is not well enough known; the fossil record is too incomplete; the periodicity is an artifact of the way Sepkoski's catalogue was compiled; and so forth. Others criticized the statistical analysis and the possible biases that can produce false patterns out of noisy data. Even *The New York Times* published editorials ridiculing the idea of periodic extinction.

In science, a regular pattern like the presumed 26-million-year extinction interval demands a plausible explanation before it can be taken seriously. Raup and Sepkoski had no clear ideas themselves, but they realized that such long periods are typically the domain of astronomers. They raised the question at a meeting devoted to the K-T extinction event, a meeting attended by many astrophysicists, who were quick to answer the challenge. The most persuasive suggestion is that the sun has a small companion star, perhaps a tenth of the sun's mass, that adopts a highly elongated orbit and passes through the Oort cloud of cometary debris every 26 million years. This star, dubbed Nemesis by one of the groups that first proposed it, disrupts the cloud and causes a shower of destructive debris to rain down on the inner planets. If this scenario is correct, the "death star" as it has come to be known is now about two light-years from Earth. Such a faint nearby star would be extremely difficult to recognize against the backdrop of millions of other brighter stars, but astronomers keep alert for a faint star that seems to move against the backdrop.

Many scientists find the periodic-extinction idea appealing, but controversy still surrounds the hypothesis and much research remains to be done. Sepkoski has been compiling a new computer compendium, this time with tens of thousands of fossil genera (the next finer taxonomic division compared to his original family list), in hope of examining the history of life at a more detailed scale. Geologists have commenced a parallel search for periodicity in large-impact events and in the iridium anomalies that they seem to produce. Astronomers continue to search for Nemesis. In Raup's words: "To some observers, the 'discovery' of the 'death star' heralds a completely new way of looking at the Earth and its life, a scientific revolution. To others, it is science gone mad."

Not infrequently, answers to one big question shed light on others. The Nemesis companion-star idea has intriguing implications for SETI, the search for extraterrestrial intelligence. Most SETI efforts have been focused on single stars, because it has been assumed that the much more common binary star systems are too chaotic to favor the evolution of life. But the Nemesis idea suggests that periodic stress caused by the sun's companion stirs the evolutionary pot. Many researchers have argued that mammals, and thus humans, could not have evolved without the demise of the dinosaurs. Earth's hypothetical companion star, therefore, might have played a key role in the appearance of intelligence on Earth. Perhaps SETI researchers should take a closer look at double star systems.

Will Another Killer Asteroid Hit the Earth?

Scientists have identified about 150 ancient impact craters in what has become an intensive global search. Some of these objects, ranging from about a half mile to hundreds of miles across, display clearly recognizable circular features. Most impacts, however, must be identified by irregular outcrops of unusual rocks, typically shattered and partially melted zones of intense deformation. The search has only begun. Based on the limited extent of present crater surveys, experts estimate that all known objects may represent only a tenth of the impact remains yet to be found. The vast majority of major Earth impacts, furthermore, have long since been eroded away, or swallowed up in subduction zones where old plate material plunges into the mantle. Throughout geological history the Earth has been bombarded by many thousands of lethal projectiles.

Whether or not mass extinctions are periodic, without any doubt Earth will be hit by another giant asteroid or comet, and another extinction event will occur. Whether this event occurs in ten million years, or next month, no one can say for certain, but it will happen—unless we do something about it. In an effort to estimate the probability of a catastrophic impact by an object greater than a mile in diameter, Clark Chapman of the Planetary Science Institute in Houston and David Morrison of NASA compiled data on the number of asteroids in Earth-crossing orbits.

In their prominent *Nature* article, which appeared in January 1994, they concluded: "There is a 1-in-10,000 chance that a large asteroid or comet will collide with the Earth during the next century, disrupting the ecosphere and killing a large fraction of the world's population." In other words, "Each typical person in the United States stands a similar chance of dying in an asteroid impact as in an airplane crash or in a flood." While the risk of a catastrophic impact is orders of magnitude less than an airplane crash, the net risk to each of us is offset by the huge loss of life predicted for a single impact.

Carl Sagan, whose articulate advocacy of space exploration fired the imagination of readers throughout the world, proposed two key strategies to avert such a disaster. The first, simpler step is to conduct a careful and ongoing survey of objects in Earth-crossing orbits. If we are successful in monitoring every dangerous rock in space, we can at least predict any risks well in advance.

The second, far more ambitious step is to establish a human presence in space. Only in this way, Sagan advised, can we insure that our species will survive a global catastrophe. As we become adept at living in space, furthermore, we might learn to deflect any threatening object long before it approaches Earth. Of course, such technology was but a small step in Sagan's ultimate dream of human exploration beyond our own solar system.

Biologists might well demur from Sagan's conclusion that space colonization will allow survival of our species. Colonization of new and isolated ecological niches is very often followed by evolution to new species. This is one reason evolution occurs rapidly and distinctively in the specialized environments found on isolated islands. More than one third of the known species in the fly genus *Drosophila* are on the Hawaiian Islands, and they most likely evolved from a single common ancestor that made its way from older land masses to the relatively young islands. Evolutionary change in humans colonizing space is likely; eventually some new species of *Homo* should emerge—*Homo martian* perhaps.

Are Humans Causing Another Period of Mass Extinction (and Will We Be Part of It)?

Of all historical extinction events, the disappearance of most large North American mammals, including mastodons, mammoths, and wooly rhinoceros, about 11,000 years ago is perhaps most sobering. Fossil bones indicate that approximately two thirds of large mammal species in North and South America died out about that time. This mass extinction differs from others in two key respects: Smaller life-forms do not appear to have been affected, and human stone weapons are found among the scavenged bones. To the list of potential causes of mass extinction we must add human activities.

According to University of Washington paleontologist Peter Ward, "We are in the middle of a mass extinction as profound as any that can be recognized in the geological record." The growing evidence is staggering. Australia, home to fifty species of large mammals just 50,000 years ago, now has only four. Hawaii has lost more than half of its approximately one hundred species of native birds, and perhaps 99 percent of its estimated one thousand species of land snails (which have been highly sought after by collectors), in the past two centuries. Humans have systematically hunted and exterminated the dodo, a dozen or so species of the Moa, the great auk, the passenger pigeon (of which there were once more than five billion in North America), and many other species. Overfishing has drastically reduced populations of wild salmon in the Northwest, blue crabs and oysters in Chesapeake Bay, and schools of fish in the northern oceans. The use of pesticides, fertilizers, detergents, and other chemicals has polluted many critical watershed areas. Ecologist Norman Meyers estimates that Brazilian species disappear at the rate of four per day, while biologist Paul Erlich fears that by the year 2000 the planet's extinction rate will be on the order of one every few minutes.

As the human population increases virtually without control, wilderness areas, especially the dwindling tracts of species-rich tropical rain forests, are being gobbled up. To these immediate causes of species extinction are added long-term effects of global warming, ozone depletion, and atmospheric pollution, all of which lead to environmental stress.

Why should we care about the disappearance of other species? After all, humans do not appear to be in any immediate danger of extinction. Three compelling reasons have been cited for preserving Earth's biodiversity. The first reason is that all species are interdependent on others as part of complex ecosystems. Loss of one species may adversely affect the whole system with consequences that are unexpected and unpredictable. Damaged ecosystems, in turn, may affect rainfall and climate, soil nutrients and erosion, pollination rates, pest control, and water quality. As more species are threatened, the risks to humans of these adverse effects increases.

The second argument to preserve species focuses on our health and well being. Humans have enjoyed untold benefits from the new foods, new drugs, and new chemicals discovered among living things. Countless natural substances, many of them of great economic value, surely remain to be discovered. Each lost species may thus be mourned as a lost opportunity for improving our lot.

Finally, many people argue for species protection on ethical and aesthetic grounds. Eminent biologists Paul Erlich and Edward Wilson state: "People have an absolute moral responsibility to protect what are our only known living companions in the universe. Human responsibility in this respect is deep, beyond measure." We cannot easily measure our loss, or our guilt, in causing a species to become extinct.

GROWING UP, GROWING OLD: HOW DO WE DEVELOP FROM A SINGLE CELL?

These are stirring days in developmental genetics; . . . There *are* glimpses of clarity—enough to see the immensity and beauty of the problem and enough to know that there is still a long and challenging journey ahead.

—Peter Lawrence, *The Making of a Fly* (1992)

The Death of a Cell

Timing is everything in the game of life.

Consider the lowly flatworm, *Caenorhabditis elegans*. As the worm develops, its sex organs swiftly adopt their complex structure. One key component of the male reproductive system is the vas deferens, a narrow tube through which sperm must travel to congregate in the cloaca prior to fertilization. This tube grows with its end blocked by a single obstructing cell that helps seal the critical connection between the vas deferens and the cloaca.

Once the connection is made an extraordinary event occurs. Triggered by as yet unknown signals from its neighbors, the blockading cell commits suicide by digesting itself. In a matter of minutes the cell disintegrates, thus opening the pathway. If the cell were to die too soon, the sex plumbing would leak; if it didn't die at all, the system would be blocked. How can a cell know when to die?

DEVELOPMENTAL BIOLOGY

A Very Complex Process

One of the greatest mysteries of life resides in the fertilized egg, the microscopic object from which everyone develops. From a single cell arises a wide variety of specialized structures composed of many different kinds of cells—more than two hundred different specialized types in humans and other vertebrates. As an embryo develops, cells must adopt exact spatial relationships in a precise time-ordered pattern. How is it possible for the genes in a solitary fertilized egg to contain all the information necessary to produce a unique individual?

This question, first explored a century ago by German biologist Wilhelm Roux, has blossomed into one of the hottest frontiers of science, with thousands of researchers engaged in the pursuit. When editors of the prestigious weekly magazine *Science* asked top developmental biologists to identify their field's most critical unanswered questions, many agreed that the central mystery remains the step-by-step process by which an egg becomes us.

Every part of our body—every limb, every organ, every cell—has its own astonishing story to tell. The hand that turns this page, the eye that scans the text, the brain's neurons that record the memory, all arose from the same single cell, the fertilized egg. As that first cell divided again and again and again, primitive structures appeared. Head, gut, legs, and heart took on their unique identities, while new generations of cells played the specialized roles of blood, bone, and brain. How does it happen?

No other question in this book could have a more complex and lengthy complete answer. To document and describe the countless individual steps that yield a single fly—the rough bristles of its legs, the regimented facets of its eye, the exquisite tracery of its wings—would require thousands of thick volumes, each richly illustrated and dense in the jargon of genetics. For a human being the volumes might number in the millions, and we are still a long way from knowing what to put in such books.

We may never know all the details of the developmental processes that sculpt our faces, our bodies, and our minds. What we can hope to learn, perhaps within a few decades, are the general principles that govern the development, life, and eventual death of all living things. In the process we may discover new ways to heal broken and diseased bodies. Someday, patients may be able to regenerate their own damaged kidneys or lungs from a single healthy cell. Victims of brain and spinal cord injuries might be coaxed into producing new neurons and nerve cells. The instructions are deeply buried within each of us; we need only learn to read them.

Switching Genes On and Off

A carefully regulated program of gene activity underlies the development of all organisms. Similarly, genes are switched on or off in the specialized cells of mature organisms. The switches are sensitive to the cell type, to the stage of development, and to environmental factors.

Every cell of an organism has the complete genome, yet cells perform remarkably specialized functions. The human pancreas, for example, is a composite of different types of cells that produce key chemicals, including the hormone insulin, and a variety of digestive enzymes. All cells in the pancreas carry all genes, but in each cell type only a limited number of specific pancreatic proteins are manufactured. Genes for the two proteins that together constitute hemoglobin, similarly, are present in every cell but are expressed only in red blood cells. The genes for these two proteins, called alpha- and beta-globin, are located on different human chromosomes, but they are switched on (in red blood cells) or off (in other cells) coordinately. When a gene is "off," no messenger RNA is made from the gene and thus no protein; the gene is silent.

Hemoglobin gene activity is actually even more exquisitely controlled than this. Different forms of both alpha- and beta-globin genes are switched on or off at different stages of development. One kind appears during the earliest stage in human embryonic development and is replaced later by a specialized fetal globin. After birth yet another hemoglobin is made. How is it possible to turn these genes on and off with such precise timing?

Gene regulation is accomplished in many ways. The process is controlled in part by the specialized structure of chromosomes, which are packages of DNA and proteins. Depending on how tightly proteins pack the DNA in a given chromosomal region, the activity of nearby genes can be stimulated or suppressed. A chromosome's proteins can fine-tune gene expression with respect to time, cell type, and environment. Cells can control the rate at which messenger RNA is produced or destroyed, the ease with which messenger RNA is transported from the nucleus to ribosomes, the ways in which proteins themselves are transported, and the rate at which newly manufactured proteins are degraded by the cell. Virtually any step in the complex process of protein synthesis can serve as a point of regulation.

Mutants—When Things Go Wrong

Developmental biology seems to the outsider a peculiar sort of science. It's almost impossible to track the genetic pathways of development when everything goes right. Even if we could freeze the sequence of events and examine every embryonic cell at every step along the way, too many processes occur simultaneously and too many genes play a role. Humans develop much too slowly, and the ethics of embryo research are too touchy, to make much progress studying our own species.

For this reason, developmental biologists, hoping to learn how humans form when everything goes right, concentrate their efforts on much simpler fast-breeding organisms in which something goes very wrong. Bizarre mutant flies with extra eyes, deformed wings, or legs sticking out of their heads provide an intellectual gold mine for these researchers. The standard research strategy involves growing countless millions of short-lived animals, most often the fruit fly *Drosophila melanogaster*, with its convenient ten- to fourteen-day life cycle. Thousands of scientists spend their entire research lives working on genetics and development of the fly, the most thoroughly studied of all complex organisms. (A simple species of flatworm, *Caenorhabditis elegans*, with exactly 959 cells in its adult body, comes in a distant second, followed by small vertebrates, including zebrafish, frogs, and mice.)

Developmental biologists produce a high yield of mutant individuals, for example by exposing breeding flies or their eggs to X rays or mutagenic chemicals. When a new individual fails to develop, or when it develops with an obvious abnormality, the research team swoops down to identify which gene has gone awry. Step by painstaking step, as critical developmental genes are identified one by one, a picture of the key processes emerges.

Getting Started: Pattern Formation

The development of every complex organism begins long before the sex act, sometimes months or years before egg and sperm are united. In humans, each female is born with a lifetime supply of eggs, ready and waiting in the ovaries. Each of these eggs must contain a suite of complex chemical messages that will guide the initial stages of embryo formation. The first clues to this process were discovered in 1905 by pioneering biologist Edwin Grant Conklin. While examining eggs of the tunicate, a small marine animal, he noticed distinct color gradients. Some portions of the egg were yellowish, others grayish or brown. As the fertilized cell began to divide inside the egg, the embryo invariably produced different tissues—muscle, gut, nerves—according to the colors. It now appears that egg proteins produced by the mother control initial development of the offspring.

Similar results have been found in flatworms. The egg's first cell division always results in a larger cell to the front and a smaller cell to the rear. If one of those two cells is removed, the next division again yields a larger and a smaller cell in the same orientation. The separation of head from tail occurs right from the start by some as yet unknown chemical signal in the egg.

The egg can't control development forever. In the flatworm, after two cell divisions (four cells total) removal of any one cell will result in grievous deformity. Evidently, from that point on the cells themselves send each other signals that guide development. The details of those signals, which may involve subtle interactions of many different chemicals, are often obscure, but one principle is clear: Each cell's role is determined by which of its genes turn on and which turn off. One set of switches produces a nerve cell, while another set generates muscle or skin. Every step is controlled by chemical signals—a unique chemical mix in a unique three-dimensional configuration; the details can be mind-numbing, involving literally thousands of genes in complexly intertwined arrays. To identify and understand all of these genes will engage biologists for many more decades. Nevertheless, the general principles of development that are beginning to emerge are remarkably simple and appear to be shared by all complex organisms.

Consider the classic case of fruit-fly development, which is elegantly summarized in Peter Lawrence's monograph *The Making of a Fly*. The fly egg features concentrations of three maternal proteins that establish the orientation of head, tail, and underside during the first cell divisions. As the elongated embryo grows, different concentrations of these three egg proteins in different regions trigger "gap" genes that cause the embryo to form a curious pattern of seven chemical bands, which ultimately yield fourteen larval segments. The 1995 Nobel Prize in Medicine and Physiology was awarded in part for key discoveries related to this segmentation process.

Two fly segments develop in place of each of the seven bands by the triggering of new genes, including "pair-rule," "segment polarity," and "hairy" (molecular biologists revel in giving newly discovered genes and their associated proteins colorful descriptive names). As the fly larva continues to develop, each of the fourteen well-defined segments is bounded by concentrations of proteins determined by the "engrailed" and "wingless" genes. These genes play off each other by controlling the expression of still more genes; wingless activates engrailed in adjacent cells, and engrailed activates other genes in an ever-expanding cascade—a symphony of genetic interactions. In this dynamic process, called pattern formation, every cell has its own unique chemical environment and expresses its own unique mix of genes.

"Working out the detailed action of all those genes, proteins, [and their interactions] will be a hard slog and often tedious," observes developmental biologist Lewis Wolpert of University College, London. And if we are just beginning to understand the development of a fly, we know next to nothing about human development: The contributions of egg proteins, the expression of the embryo's genes, the development of patterns that will become major body parts are but poorly known. Even so, some biologists suspect that the basic principles of development are well in hand. In all developing organisms cells, at the direction of genes, release chemicals that control the destiny of their neighbors. What biologists learn about fly development is thus likely to be directly applicable to humans.

The Homeobox

Flies, worms, fish, and people rely on remarkably similar body plans. All of these organisms, along with most other animals, have a front, a rear, and a long gut extending from one end to the other. Might the developmental process be similar in all of them?

In 1984, biologists working with (what else?) the fruit fly noticed that a grotesque mutation that placed fully formed legs on the head in place of antennas was tied to a defect in a small DNA segment of a gene. This segment, called the homeobox, defines a modest protein region of sixty amino acids that itself binds neatly to a specific DNA segment, thereby regulating genes that are crucial to pattern formation, particularly recurrent patterns such as arthropod segments, fly eye facets, and mammal vertebrae.

Nearly identical homeobox segments appear more than a hundred times on the fly genome, each segment being associated with a different key developmental gene. These homeoboxes and their associated genes cluster along two giant chromosomal sequences, each a staggering quarter million base pairs long. Furthermore, the physical order of these developmental genes along the chromosomes is the same as the front-to-back order of the segments in the fly embryo. When a homeobox is broken, the fly's assembly instructions are somehow scrambled.

That discovery was interesting enough, but what really grabbed the attention of developmental biologists was the subsequent discovery of nearly identical homeobox segments in all sorts of organisms, from worms to humans. Moreover, the order of the genes in clusters of homeobox-containing genes again mirrors the order of regions along the front-to-back axis of the organism in which they are expressed. The exact same genetic switch now appears to operate in countless ways, triggering packets of genetic instructions that make all manner of anatomical structures, from teeth to toes. Some scientists now suspect that defects in these homeobox control sequences may be responsible for many birth defects and spontaneous abortions in humans.

The implications of this discovery are vast. For a time it seemed as if every step of the developmental process was a special case, dependent on its own unique blend of genetic signals. Discovery of the universal role of the homeobox confirms what many biologists had hoped, that general principles underlie the details, and that exhaustive research on fruit flies and other animals has direct bearing on human development. Both the general principles and many of the specific steps observed in fly development shed light on the much more complex and inaccessible process in humans.

Indeed, a growing body of evidence suggests that many genes crucial to the fly play equally important roles in mammal embryonic development. A gene critical to heart development in the fly, for example, causes heart defects when mutated in mice. A defective wing-forming gene in flies causes breast cancer in mice, while yet another gene leads to malformed eyes in both flies and humans.

These discoveries provide new insight to the process of genetic evolution. The homeobox is a shared genetic heritage that must date back at least half a billion years, to a time when the evolutionary paths of worms and vertebrates diverged. The homeobox mechanism proved so efficient that it has been preserved as a genetic living fossil—yet another piece of evidence for the common ancestry of all animal life on Earth.

How Do the Sexes Develop?

At some key point in the development of an embryo a very important decision must be made. Will it be a boy or a girl? Development biologists are fascinated by that question, both for its own sake and because it may serve as a model for many other decision-making processes that take place during the growth of an individual. Biologists turned to the fruit fly and flatworm for deciphering genetic details of the process. But human patients with rare genetic disorders in which males and females display physical characteristics of the opposite sex or have unusual numbers of chromosomes provide special insight on sexual development. Human patients are especially important because, unlike many other developmental processes, those that trigger the choice of sex may be quite different from animal to animal.

The majority of animals have distinct chromosomes that determine sex. Females usually have more copies of a chromosome called X than do males. In mammals, normal females have two X chromosomes and normal males one X and one Y. Sex is determined by the presence or absence of special genes on the male Y chromosome; mutations in those genes result in XY individuals that are nevertheless female. In flies and worms it is the ratio of X chromosomes to nonsex chromosomes that makes the difference.

Researchers engaged in this research tend to focus on mutations that cause trouble in the development of either males or females. This ongoing work indicates that only a few specific genes in each organism regulate male or female development; once these genes are turned on, male- or female-specific proteins are produced, and an embryo takes one of two very different paths. Curiously, those genes appear to be completely different in flies, worms, and mammals. The mystery of how sex-control genes evolved independently several times remains a hot research topic.

Sexual organisms face an intriguing challenge because of the difference in number of X chromosomes in males and females. Each X chromosome carries many vital genes that have nothing to do with gender—genes required by both sexes. How can a male, with only one copy of X, be sure that enough X-chromosome proteins are produced? How can a female with two copies be sure that she doesn't get too much? These questions fascinate molecular biologists because they represent a well-defined example of the much more general problem of gene regulation: How does a cell control how much of a particular protein to produce?

This dilemma deepened when geneticists realized that organisms have evolved different strategies to compensate for the sex's unequal shares of X chromosomes. At least three different strategies are now recognized—in flies, worms, and mammals. In all three cases the expression of X chromosome genes is increased or decreased by changes in the structure of special proteins that bind to DNA and regulate gene expression.

In flies, the breakthrough came in the 1970s, when geneticists realized that some mutations killed only developing males, while others killed only developing females. Genetic and biochemical experiments led to an understanding that one gene called sex-lethal is a key. Not only is the protein coded by sex-lethal important for determining which sex a fly will be, but in males it also sets off a cascade of other genes that results in a doubling of the level at which X chromosome genes produce proteins. Thus, males with one X chromosome and females with two express the same amounts of X chromosome proteins.

Flatworms have evolved the opposite strategy; the activity of the female's two X chromosomes is halved to match the male's output. When a master suite of female worm genes is activated, five additional genes produce proteins that bind to her X chromosomes and reduce gene action. Female mammals use yet another dosage compensation strategy. Special regulatory proteins inactivate the genes on only one of the two available X chromosomes, thus exactly halving the female's X chromosome protein output. Remarkably, the choice as to which X chromosome is blocked appears to be completely random.

Gene regulation must maintain an exquisite balance in every cell throughout life. When these crucial mechanisms go awry, the results are often catastrophic.

WHAT DO WE KNOW ABOUT AGING AND DEATH?

No aspects of our physical existence have more power over our hopes, aspirations, sorrows, and fears than aging and death. But the fact is that aging and death are not well understood, and relatively few scientists are actively addressing questions relating to them. We can ask, "Must we die?" but no one yet has a clear idea how to answer that question.

We live through the dimension of time. Birth, adolescence, middle age, and death describe temporal stages of our passage. But what is the role of time at the level of life's universal unit, the cell? Do cells tell time? Do they age? What causes death at the cellular level?

Long-lived organisms

A hallmark of modern medicine is the dramatic prolongation of human life. Three centuries ago, our average life expectancy was about thirty-nine years. In 1900, sixty-five was considered a ripe old age. Life expectancies are now in the seventies, and it's not difficult to imagine a future in which hundredth birthdays are commonplace. Is there a limit to human life spans?

Unusually long-lived organisms may provide some clues. About the time that Roger Williams settled in Rhode Island more than three hundred years ago, tiny crustaceans called copepods thrived in local ponds, just as they do today. Recently a team of biologists from Cornell University found that antique copepod eggs, long buried in pond mud, can be brought to life when placed in warm water and light. Experts anticipate that even older eggs may be found in ongoing investigations. These findings lend some credibility to persistent but still-unconfirmed reports of viable seeds estimated to be thousands of years old from Egyptian tombs and Japanese bogs.

Whatever their age, these cells are modern compared to the reported antiquity of some geologically preserved strains of bacteria. In one eye-opening (and hotly debated) series of experiments, biologists Raul Cano and Monica Borucki of California State Polytechnic University extracted dormant bacterial spores from a Dominican stingless bee that they claim became trapped in amber tens of millions of years ago. Rumors also persist of halobacteria that some scientists suggest were locked in salt domes and appear to come back to life after a quarter *billion* years of dormancy. When placed under stress, these types of bacteria secrete a thick outer shell, their metabolism stops, and their lone chromosome becomes dehydrated. Such protective adaptations, combined with the hermetic seal provided by amber or salt, may have preserved the cell until it was exposed to a more friendly, water- and nutrient-rich laboratory environment. If these organisms can remain viable for so long, is it possible to lengthen humans lives?

Cellular Clocks

The ability of cells to change or act at the "right time" differs fundamentally from the gradual phenomenon of aging. Many chemical processes that define the functions of a cell must somehow tell time. This timekeeping regulates an orderly sequence of events at a cellular level that can, for example, take a single cell through its development to a full multicellular organism, or an organism through its stages of development. A cactus blooms, a tadpole metamorphoses into a frog, and a human child passes into puberty through chemical changes at the cellular level.

For a cell to change over time, it must receive a stimulus that causes one or more specific genes to be read. This reading takes time, and each gene, depending upon the stimulus, is read at different rates. Furthermore, the number of times a gene is read, whether by the minute, the hour, or the day, varies. Proteins, the products of genes, are thus manufactured at different rates by different cells. Ultimately, any change in a cell must result from variations in the rates of protein production and destruction.

For the first time, scientists studying life at the molecular level are catching tantalizing glimpses of the mechanisms by which cells change over time. Carefully orchestrated changes in the rate of protein creation and destruction are driven by the ebb and flow of internal and external signals that turn up or down the reading of genes. Molecular biologists now struggle to understand how this precise control is achieved, and how midstream corrections are made in the intricate choreography. The problem is enormous, because each cell must orchestrate the varying expression of tens of thousands of genes.

Most genes in a cell are silent until a signal arrives to turn them on, while other genes are active until a signal arrives to turn them off. Some genes may be switched on and off many times, while others, once triggered, become permanently on or off, even after the initiating signal is gone. The products of newly expressed genes alter cells and can signal new patterns of gene expression, either in the same cell or different cells. Such "cascades" of gene expression are the topic of intense exploration, for in their description lies the temporal patterns of change that cause the development of all complex, multicellular organisms. Observations of such phenomena have shown scientists how to begin to think about changes in a cell over time.

Cascades of gene expression provide a mechanism that can explain developmental changes over time, but can organisms tell time? On the one hand, we know of an extraordinary array of what seem to be internal clocks that produce rhythms. Secretions of hormones, for example, often follow multiple rhythms from rapid pulses of a heartbeat to daily patterns of eating and monthly reproductive cycles. These rhythms, while often mysterious in detail, may result from feedback loops of gene expression: The protein product of gene A reaches a concentration that triggers expression of gene B, gene B's protein triggers gene C, and so forth back to gene A. Some of these cycles can be set and reset, for example by the effect of the sun on light-sensitive proteins. Your body's daily cycle resets each time you experience jet lag. Such rhythms might be interpreted as biological clocks, because a cycle, especially a regular one, can be used to tell time.

The Ultimate Clock

On the other hand, a more fundamental question about biological time relates not to rhythms, but to history—the question of aging and death. Does a cell or an organism have a way of counting events so that it "knows" not where it is in a biological cycle, but *when* it is in relative time, say from birth or death?

As we are painfully aware from everyday experience, organisms age. But why should this be so? What is aging? At one level aging can represent the accumulation of damage—rust from the weather, scratches and dents from use and misuse, and the simple physical wearing out of parts. Organisms are constantly subject to this kind of damage: A football player's knees and a violinist's shoulder wear out with use and stress. The accumulation of toxins or cholesterol-filled plaques in blood vessels gradually compromises function and increases the probability of a catastrophic event such as a heart attack. In that sense, we are like a machine that can wear out. But does accumulated damage alone explain aging and death? And what about cells—do they age?

The question of cell aging is one of the most intriguing in biology. Virtually all cellular components can be damaged by normal physical and chemical stresses. Indeed, damage to cellular molecules is not a rare event, like a car accident, but rather a constant fact of cellular life. A significant difference between a car and a cell, however, is that the cell is able to repair its damage. Virtually all cellular components are made by the cell, so new components can be made to replace old damaged ones.

All instructions for manufacturing cellular components are coded by DNA, so what happens when DNA is damaged? For the first decade or so after the structure of DNA was discovered in 1953, relatively little attention was paid to the issue of damage and repair of DNA. When scientists involved with the discovery of DNA structure have been asked about this oversight, they often respond by pointing out that the structure and meaning of the DNA molecule is so beautiful, so perfectly suited to carry and pass on life's message through billions of years of evolution, that they just assumed DNA was somehow protected from damage. Yet scientists already realized that some external influences, like X rays and some chemicals, could damage DNA and cause mutations.

The question of DNA damage and repair now drives a thriving field of molecular biology. It turns out that DNA is a very vulnerable molecule, by some estimates suffering over 10,000 structural "errors" per cell per day. Some of these errors are the result of broken interatomic bonds, due to damaging interactions with water, oxygen products, ultraviolet light, or any of thousands of cellular or environmental chemicals. This high rate of damage is one reason many molecular biologists doubt reports of dinosaur DNA and other ancient genetic material.

Other DNA errors result from mistakes made in copying the DNA in the process of cell growth and division. The enzymes that copy DNA insert the wrong base about once every 100,000 bases. These enzymes can also monitor their own mistakes and correct 999 out of every 1000 errors. While the result is an incredibly efficient system, about thirty mistakes per total genome escape correction during every division of a human cell. Despite this knowledge, we still don't know whether the accumulated errors in DNA allow the cell to tell time and age.

In the late 1950s Leonard Hayflick performed a crucial experiment that shed light on the nature of cellular aging. His breakthrough depended on a method for growing isolated animal and human cells in dishes, a technique called tissue culture. An enormous amount of trial-and-error work led to complex recipes for nutrients in which cells could be nourished, allowing them to grow and divide in culture. Remarkably, in such a growth medium an individual human cell will adopt an amoebalike shape and multiply like a single-celled organism.

In these experiments Hayflick and his coworkers found that no matter how the nutrients were manipulated, normal cells would divide only a certain number of times—typically about fifty—and then they would die. Interestingly, cells taken from young animals divided more times before dying than those taken from older animals. Different types of cells divide at different rates, but it is not time that these cells seem to keep track of, but rather the number of cell divisions. A striking exception to this rule is cancer cells, which are for all intents and purposes immortal in culture. Thus, it appears that most normal cells have a clock that can tell time through history by counting the number of divisions.

Hayflick originally proposed that this phenomenon represents an inexorable accumulation of errors, but now scientists are beginning to ask whether a more specific molecular mechanism is behind the workings of this biological clock. The articulation of this new question illustrates the heart of scientific discovery. The new search for a specific, definable molecular clocklike process replaces the previous vague and fuzzy idea of accumulated damage. The new question creates an intellectual doorway leading to unanticipated discoveries.

We don't yet know what the molecular clock is, but one intriguing hypothesis involves special structures called telomeres. The core of each chromosome is a long DNA helix containing 50 to 250 million base pairs. At the ends of each chromosome are found special sequences of nucleotide bases, the telomere, which are repeated hundreds to thousands of times. Most linear chromosomes, from those in single-cell yeast to humans, possess these terminal structures. The universal mechanism of copying DNA, an event that accompanies each cell division, is incapable of copying all the terminal repeats. Thus, with each division of a cell, the number of terminal repeats decreases. Perhaps each shortening episode, each cell division, is a tick of the cellular clock. Perhaps when the chromosome ends become too short, the cell dies.

A few types of cells do not experience permanent terminal shortening, thanks to an enzyme called telomerase that can rebuild the ends, or telomeres. This enzyme appears to be characteristic of many cancer cells, as well as germ cells (cells that give rise to sperm and egg). These two types of cells are in effect immortal. Ordinary cells that are mortal, on the other hand, appear to have low levels of telomerase. Whether the shortening of telomeres is the cellular Father Time remains to be proven, but the search continues.

Cell Death

We all die, and that death generally results from the failure of one or more critical organs as a result of disease or trauma. Similarly, a cell in isolation can be readily killed by heat, physical forces, or viral infection. Most cell death, however, is not murder by outside forces, but rather a form of suicide. Given the appropriate signals, a cell may spontaneously burst into pieces that are easily digested or disposed of, usually in a matter of minutes. In fact, the orderly development of a new organism depends on the formation of some cells that are determined by genetic programs to die before the individual reaches maturity.

Programmed cell death first comes into play during the sculpting of a growing embryo. Selective cell death creates embryonic pits, cavities, and tubes that will become major structures of the body. The cells that form the weblike tissues between developing digits, for example, all die at a certain time in development to leave isolated fingers and toes. In the brain, cells die at enormous rates, leaving only cells that have established productive, interacting networks. Programmed cell death eliminates cells of the immune system that recognize and would attack our own molecules. Then, throughout life, programmed cell death serves in the crucial roles of controlling numbers of cells and helping to excise damaged or defective cells.

Today, researchers are discovering new examples of programmed cell death at an accelerated pace. The molecules involved in cell death are being identified, while certain human diseases are being traced not to the presence of cell death, but to its absence. These discoveries naturally lead us to ask whether programmed cell death ultimately contributes to aging and death. If we escaped accidents and specific disease, could we live forever?

The answer now appears to be no. How much of this inherent limit in human life span is due to accumulated damage and how much to the intrinsic life span of cells is not clear. Perhaps a suicide instruction is triggered by a clock that counts cell divisions. If the telomere is reduced below a critical length, for example, it may trigger a specific pathway of programmed cell death. Perhaps the cell can decide "enough, it is time to die."

To explain cellular aging and death, the issue of accumulated damage has been contrasted with that of voluntary suicide, as if these issues were mutually exclusive and disconnected phenomena. They are, in fact, intimately linked, and nowhere is this connection more clear than in our developing knowledge of cancer.

Cancer: Life, Death, Damage, and Repair

The human body, crafted from perhaps a hundred trillion cells, must maintain an exquisite balance. Regulatory processes that must be encoded in our genes but are not yet well understood orchestrate the death of old cells, while new cells arise to take their place. Other modulating genetic processes regulate the rate at which cells divide to form new cells. A tumor occurs when defects in the genetic machinery cause a cell to divide again and again in a runaway fashion. The great irony of cancer is that it results from the *failure* of cells to die.

Only recently have scientists begun to understand what the fundamental difference is between a normal cell and a cancer cell. A normal cell ages and dies, while a cancer cell is immortal because its clock is either turned off, constantly reset, or just ignored. Here the interplay between damage and death comes strongly into play: The death of a cell is a choice driven largely by an extraordinary property of cells—the ability to recognize and assess damage.

The inevitable drive of life is to copy DNA and distribute it to the daughter cells. Life on this planet, in all its incredible variety, represents many different solutions to the passage of DNA copies from one generation to the next and from one cell to its daughters. It has been said that the chicken is the egg's way of making another egg. More accurately, the chicken and egg together are the astonishing solution to the problem of passing chicken DNA from one generation to the next. The cycle of life is an ordered process of precisely duplicating the DNA, accurately separating the two copies in space, and carefully dividing the cell so that each of the two new cells receives an identical set of DNA molecules.

This cellular cycle can be divided into a series of ordered steps separated by discrete transitions. As a result of research on defective cells that fail to make it through the entire division sequence, biologists are beginning to document the detailed chemistry of these transitions. In the course of these momentous discoveries, it has become clear that each transition awaits both external and internal signals before proceeding.

The internal signals, which tell the cell that it's okay to proceed with the next step, are particularly interesting. Does the cell have enough nutrients, and has it grown enough to initiate DNA duplication? Have there been mistakes in DNA duplication? Has the DNA been duplicated fully and only once? Are the chromosomes correctly segregated? Is the cell oriented correctly to partition the chromosomes? At each transition the cell stops to check whether the phase it has just completed has been successful and whether the cycle of DNA duplication and partition has been accomplished with the extraordinary fidelity that life demands.

The several stopping points along the path to cell division, called checkpoints, serve as essential guardians of life. Details of these cellular inspection mechanisms remain largely unknown, but we do know that if the checkpoint machinery spots a defect, then the cycle is stopped and one of two things happens: Either the defect is repaired or the cell commits itself to die via programmed cell death. Thus damage and programmed cell death are intimately connected, as long as the checkpoint machinery is intact.

Cancer, a disease of continued unsupervised growth and genomic instability, occurs when this fundamental guardianship fails. The cell cycle continues unchecked, often despite profound damage to the DNA and chromosomes. By ignoring chromosomal damage, additional harmful changes accumulate, releasing the cancer cell from even more controls over its normal behavior. The result is a terribly damaged cell that instead of dying as the result of damage, thrives because it will not or cannot commit suicide in response to that damage. The very existence of immortal and irrepressible cancer cells suggests that damage to DNA per se does not cause a cell to age. Indeed, cells can thrive despite deep and extensive genetic defects. Paradoxically, cancer tells us that much, perhaps most of aging and death is not the accumulation of damage, but rather the cell's response to it.

Over the past several years, scientists have succeeded in identifying several specific genes whose protein products affect checkpoint pathways. Mutations in several of these genes appear to be essential for the development of cancer. In general, no single genetic abnormality is sufficient for a normal cell to become a cancer cell; rather, a handful of such genetic abnormalities are required. Many of these DNA changes are acquired or accelerated by exposure to chemicals that damage DNA, such as are found in cigarette smoke. But the probability that any single cell will acquire the multiple gene abnormalities required to change the cell from normal to tumor behavior is relatively low. This situation is quite different for some individuals. Up to ten percent of the over 1,100,000 cancer patients newly diagnosed each year in the United States has inherited an abnormality in one of these genes. Because these individuals already carry what is called the first genetic hit on the road to cancer in all of their cells, the probability that any single cell will accumulate the multiple DNA cancer hits needed to make a tumor is much higher than normal. These individuals can carry lifetime risks of developing cancer of eighty to ninety percent.

All aspects of the cell cycle and checkpoint pathways may be affected by defective genes that contribute to cancer development. Some of these genes, called oncogenes, act as accelerators that drive the cycle continuously. In tumor cells, these genes (or the proteins they encode) are generally inappropriately turned on and push the cell through checkpoints that the cell would otherwise not pass. Other genes, called tumor suppressor genes, normally act as brakes that block the progress through cell cycle transitions and respond to damage by turning on either the stop-repair or cell-death pathways. The normal functions of these genes are often lost in cancer. A third type of cancer-susceptibility gene encodes products directly involved with the DNA repair process or the proteins that participate in programmed cell death.

Dr. Paul A. Marks of the Memorial Sloan-Kettering Cancer Center in New York predicts that "over the next few years, for every major cancer—breast, prostate, colon, lung, and ovarian, as well as many of the rarer forms . . . —[a damaged or mutated] gene or genes will be identified whose presence increases the risk for the specific cancer or cancers."

The growing recognition of a genetic basis for cancer points to new approaches for its treatment. If physicians can determine which genes are defective and at which checkpoints damage is ignored or poorly repaired, then they may be able to develop drugs that target these potential vulnerabilities. The future of research to ameliorate or cure cancer, a disease that will strike one in three of us over the course of a

lifetime, lies in the identification and understanding of cancer susceptibility genes and the cellular pathways of their products.

The behavior of each cancer is probably a manifestation of its specific pattern of genetic errors. This pattern will lead to predictions of a
tumor's aggressiveness, its likeliness to spread, and its response to therapy. Someday that therapy will be tailored to the specific genetic fingerprint of each person's cancer. As we discover and identify a growing list
of inherited and acquired cancer-susceptibility genes, we will be able to
determine the best preventative treatments for individuals at risk.

As scientists delve further into the details of human development, aging, and death, we begin to glimpse how much the unanswered questions related to human health are not and cannot be confined to science alone. Scientists along with the rest of society must ask what we
should do with information about a genetic predisposition to disease.
How will people cope with such information? Who should have access
to such information? How will it be used, and how will we protect
individuals from genetic discrimination?

As we learn to read and interpret our genetic scripts, we must recognize that each and every one of us brings into this world a set of genetic
predispositions that we did not choose. We must ask how strictly deter-
minative these scripts are, and how their outcomes can be modified by
behavior, diet, or medicine. These social and ethical issues remain as
compelling a set of unanswered questions in biology as the pathways of
cell development and death.

THE
HUMAN BRAIN:
WHAT ARE
THE PHYSICAL
ORIGINS OF
MEMORIES?

Every time you walk away from an encounter, your brain has been
altered, sometimes permanently. The obvious but disturbing truth
is that people can impose these changes against your will.
Someone can say something—an insult, a humiliation—and you
carry it with you as long as you live. The memory is physically
lodged inside you like a shard of glass healed inside a wound.
—George Johnson, *In the Palaces of Memory* (1991)

The Bizarre Case of Phineas Gage

On the 13th of September 1848, railroad construction foreman Phineas
P. Gage suffered one of the most gruesome injuries imaginable, and in
the process became a walking experiment in brain science. The twenty-
five-year-old Gage was responsible for setting explosive charges during
the excavation of rock cuts for the Rutland and Burlington Railroad,
which passed through difficult, mountainous Vermont terrain. The tedious process required repeatedly drilling deep vertical holes into solid
rock, filling the holes with explosives, inserting a fuse, covering the
black powder with a protective sand barrier, and then carefully tamping
down the entire assembly with an iron retaining rod.

On that fateful late summer day Gage inserted a one-inch-diameter
three-and-a-half-foot-long iron rod and began to tamp down the explosive charge. Tragically, he had forgotten the protective sand barrier. As
he leaned over the hole, the black powder exploded prematurely, rock-
eting the slender rod up through Gage's left cheek and eye socket and
out the top of his head, leaving a gaping hole in his brain.

Miraculously, Gage survived. Within moments of the incident he
was alert and, as his horrified coworkers looked on, he was able to talk
and move about. His astonishing recovery, with little loss of speech,
motor control, or learning and reasoning ability, astonished the medical
community. But the accident did not leave Phineas Gage unaffected.
The once diligent and courteous workman, admired by his peers for his
intelligence and gracious demeanor, became an unreliable and forgetful
employee, prone to disturbing antisocial behavior and bursts of offen-
sive language.

Gage's irreversible transformation raised unsettling questions about the relationship between a man's brain and his soul. The idea that behavior and social responsibility somehow reside in the physical structures of the brain's frontal lobe was loathsome to many theologians and other scholars of the mid-nineteenth century. Yet the case of Phineas Gage, along with those of subsequent victims of brain trauma, painted a convincing picture of the brain as the seat of emotion and personality.

For a century and a half, profound mysteries have remained unsolved. How can a grapefruit-sized mass of interconnected brain cells carry the essence of human creativity, curiosity, emotion, and reason? How can such an organ store a lifetime of experience and learning in the memories that each of us holds so dear?

THE SCIENCE
OF MEMORY

What is the most fundamental unanswered question about the brain? Many of science's deepest thinkers, including Nobel prize–winners Francis Crick and Gerald Edelman, and mathematician Roger Penrose, would say without hesitation: "What is consciousness?" Crick, who defines consciousness operationally as "attention and short term memory," has called for an intensified research effort in his book *The Astonishing Hypothesis*. Crick argues that "your joys and your sorrows, your memories and your ambitions, your sense of personal identity and free will, are in fact no more than the behavior of a vast assembly of nerve cells and their associated molecules." Experimental studies of visual awareness, he concludes, are the key to the rigorous study of self-awareness.

Many researchers, including Stanford computer scientist Terry Winograd and the late physicist Richard Feynman, are not persuaded. They despair of finding a concrete physiological definition of consciousness, much less an unambiguous experimental protocol for its study. They contend that since a clear research strategy is lacking, consciousness must for the time being lie outside the domain of science. Indeed, most unanswered questions about human thought seem to fall somewhere in the nebulous realm between philosophy and science. What is an idea? What is an emotion? What does it mean to be curious or to know something? It's hard to see how these abstract questions can be reduced in any neat way to a collective property of brain tissues, nor is it obvious how to make the giant leap from the concept of thought to a reproducible experiment in the lab. Many questions about consciousness seem to defy a mechanistic explanation (though the same was once said of the universe, human development, and heredity).

Memories are different from consciousness; they are more tangible and tightly defined. At one level memories are a kind of information that can be stored, recalled, altered, or deleted—all familiar tasks in the computer age. It's conceivable that each memory is stored in the brain as a molecule or set of molecules that carries a message. Alternatively, memories might be hard-wired into networks of brain cells, or maybe they consist of electrical potentials that pervade the whole brain. Whatever the nature of memories, we can hold out the hope that answers will yield to clever and persistent study.

There's another reason that the quest to understand memory holds a central position in the study of the human brain. Awareness, perception, and thinking depend on receiving information through our senses and analyzing that information in the context of learned patterns of experience—patterns recorded as memories. We cannot be self-aware without a remembered context of existence and personal history. Understanding the physical basis of memories, therefore, is an essential step to knowing what it is to be human—to be conscious of memories.

Searching for the Structure of Memories

The reductionist view of memories, adopted by many neurobiologists since the first decades of this century, is that remembering takes place at the level of brain cells and their interconnections. Memories, according to this view, are physical entities impressed upon brain tissues like a message carved in clay. They must be formed of atoms and assembled into three-dimensional constructs. Each remembered face, experience, sensation, and melody is preserved in such a trace, a tangible structure given the name engram.

The electronics revolution demonstrated how all manner of information, from numbers and graphs to pictures and sound, can be stored, shifted, and retrieved in the form of electromagnetic patterns. While the brain is vastly different from a computer, new digital technologies suggested a physical basis for the engram and strengthened the hope that memories can be studied with quantitative rigor. And as biologists searched for the hardware, philosophers had a field day debating whether all engrams are learned, or if perhaps humans are born hard-wired with some deep-seated, shared cultural memories.

Memory researchers looking for engrams adopted a variety of straightforward strategies. One group led by Karl Lashley of Harvard University taught rats how to run a maze—they watched as the rat seemed to create a new engram. Then they sliced out ever larger pieces of brain to see which clump of cells held the memory. Lashley's team was surprised to find that while any brain loss tended to impair a trained rat's maze-running abilities, all surgically altered rats retained some of their maze memories. Their conclusion: Engrams are whole-brain phenomena, not associated with any specific brain location.

A very different picture of the engram emerged from research by Canadian neurosurgeon Wilder Penfield, who specialized in treating severe epileptics. Hoping to locate and treat the physical sources of the seizures, Penfield cut open patients' skulls and performed a series of operations on exposed brains. He prodded different regions with an electrode, while the patients, who had to be kept awake during the procedure, responded to the stimuli. Occasionally a touch would evoke a completely unexpected response—a vivid memory of a childhood in-cident, a long-forgotten pungent smell, or a poignant fragment of mu-sic would flood the patient's consciousness. Touch the same spot again, and the same memory would come rushing back. Penfield believed that he had tapped into specific engrams, and he concluded that memories are highly localized in the brain.

So which is it? Are memories localized or spread throughout the brain? For a brief time, in the late 1960s and 1970s, researchers thought they had an answer that satisfied both camps. Neurobiologist Allan Jacobson taught a group of rats to associate a flashing light with food. He then extracted RNA from the rats' brains, injected it into untrained rats, and reported that the memory had been transferred. The un-trained rats seemed to respond to the light. Other researchers obtained similar results with flatworms that had been "taught" to avoid light:

untrained worms, when fed chopped-up trained worms, were also observed to avoid light.

The idea that memories are carried in a coded molecule had tremendous appeal at a time when the genetic code of DNA was just being deciphered. Molecules could be dispersed throughout the brain, in accord with Lashley's work on rat brains, but memories could be triggered locally as observed with Penfield's epileptic patients. And the possibility that memories and learning could be injected like a vaccine suggested to some a brave new world of pharmaceutical education.

The molecular engram craze has passed. Subsequent efforts by many research teams failed to duplicate the original studies, nor was conclusive evidence of a memory molecule found. Overeager researchers apparently saw in the behavior of their rats and worms what they wanted to see. Neurobiologists now look to nerve cells and their interconnections to locate the structures of memory.

Neurons, Dendrites, and Synapses

Brain research is an endlessly challenging game, full of compromises. Ultimately, we want to understand human brain functions, but the human nervous system is complex beyond imagining and difficult to study with rigorous controls. The nervous systems of simple animals, on the other hand, are easy to manipulate and document, but they lack the very complexity that makes humans unique. Which system to study? A life in the laboratory can so easily be consumed taking the wrong tack. As often happens in science, competing teams of researchers are approaching the problem from many different directions.

One assumption shared by all neurobiologists is the fundamental importance of neurons and their interconnections. Neurons, the complexly branching nerve cells that form animal brains, may be visualized as something like a tree. Electrical impulses come in from many "roots," or dendrites. These signals are processed and integrated by the main body of the neuron and new signals are sent out along a branch, or axon, which connects to the dendrites of other cells. Each junction between an axon of one neuron and a dendrite of another is called a synapse.

Synapses are only rarely direct electrical connections between cells.

Most often the axon and dendrite don't touch. Instead, the sending neuron releases special chemicals into the tiny space at the junction. The interaction of the chemicals with the receiving neuron initiates the next electrical impulse. These dynamic links between one neuron and the next are thus electrochemical staging areas, rich in ions and molecules that encourage or thwart the passage of signals from cell to cell. When a nerve of an eye, the skin, or other sensory organ is stimulated, either a nerve impulse speeds to the brain, or it doesn't. Individual nerve impulses come in only two varieties: "on" or "off," like a light switch. But once that signal enters the brain, it quickly cascades through a complex branching pathway of many neurons and across many synapses, triggering other nerve impulses that may stimulate, for example, the activity of muscles, organs, or glands.

Synapses, on the other hand, behave as both a switch and a volume control. The narrow space between each dendrite and axon can be flooded with chemicals that enhance or inhibit the transmission of a signal. The reaction of the receiving neuron depends on the concentration of chemicals, and different synapses have different threshold values and respond differently. Whatever brain activity neurobiologists wish to study, synapses are where the action is centered.

Simple Systems

Some simple animals such as snails and flatworms have only a few hundred neurons, with a few thousand synapses. In such cases it is possible to produce a detailed map of the entire neural network and study these cells one by one. (The human brain, by contrast, contains perhaps a hundred billion neural cells, with a hundred trillion synapses.) Ongoing laboratory experiments on both simple invertebrates and complex mammals are now providing insight on how memories form.

The sea snail *Aplysia* is a simple creature with limited learning capacity. Nevertheless, neurobiologist Eric Kandel and his colleagues at Columbia University have taught *Aplysia* a few tricks. The snail normally responds to a tap by withdrawing its gill, but this reflex can be altered with appropriate conditioning. Repeated tapping desensitizes the snail and lessens the strength of the withdrawal reflex, while tapping synchronized with a mild electric shock to the snail's tail sensitizes the animal and increases the reaction speed. In either case *Aplysia* learns a new response and its neural network is altered. By analyzing what was happening at individual synapses, Kandel concluded that learning alters the concentration of neurotransmitter chemicals at appropriate synapses—it changes the volume control. By using chemicals to strengthen or weaken a network of synapses, memories are imprinted on the snail's tiny brain.

An alternative view is held by Gary Lynch, a neurobiologist at the University of California's Irvine campus, who believes he has evidence that concentrations of neurotransmitters remain more or less constant, but that the number of synapses increases as memories form. Under the appropriate conditions, brain cells actually sprout new axons and dendrites. To some researchers this idea is heresy: The adult brain is not supposed to change.

To test his radical hypothesis, Lynch studies the neurons of rats. In one sequence of experiments he destroyed a small portion of the living rat brain, trained the rat, and searched for evidence of new synapse formation. Lynch's team made headlines in 1971 when they confirmed the growth of new neural connections, but that was only a first step. What process could trigger the growth? Years of research pointed to a variety of plausible chemical agents that physically alter the shape of a nerve cell, somewhat analogous to the way a round red blood cell becomes spiky during clotting. New spikes become new dendrites and, eventually, new synapses.

Lynch and Kandel may both be correct; memories may represent a complex interaction of new synapses *and* varying synaptic potentials. But whatever the origin, our memory processes are profoundly different from those of a digital computer. The human brain isn't just a hundred-trillion-bit binary computer, nor can it be modeled by analogy to a PC. Rather, it is an intricate interconnection of a hundred trillion volume controls. *And*, if Lynch is correct, it can make new connections if it needs to.

Whatever controls memory—long-term chemical alteration of synapses, the growth of new neural connections, or some combination of these and other physical processes—neurobiologists are still far from a detailed understanding of memory at the molecular and cellular scale.

But whatever memory might be, it's very different from anything you can buy at CompUSA.

WHAT IS THE
ARCHITECTURE OF MEMORY?

In many disciplines, including particle physics, chemistry, and genetics, collective properties are studied and understood to some degree by investigating interactions at the level of individual components. Quarks and leptons combine to make atoms, atoms bond to form molecules, and molecules cluster to fabricate cells. But the huge gap that separates the study of brain cells from their intangible collective properties—the nuances of emotion, reasoning, and personality—seems vastly greater. A memory can't be weighed, nor can a thought be X-rayed. It is difficult to imagine how an understanding of brain behavior could ever be formulated from simple reductionist logic.

As if the difficulty of studying a memory weren't imposing enough, the place where the action takes place—the living human brain—is all but inaccessible to direct study. Researchers are profoundly limited for both technical and ethical reasons in the experiments that they can perform. Except in the unusual instances of victims like Phineas Gage, who survived localized brain damage, neurobiologists must be extremely conservative in how they probe human brains. To be sure, ever more refined external sensing techniques, which measure the intensity of the brain's electromagnetic fields or the distribution of radioactive tracers, provide tantalizing glimpses into the workings of the living brain, but these procedures do not yet have the speed or resolution to document neural pathways of an idea or memory.

Given such limitations, scholars seeking to understand thought have flung a wide net. At one end of the research spectrum, psychologists investigate relationships between the physical brain and human behavior, focusing special attention on the characteristics of individuals with damaged or altered brain structures. At the opposite extreme, philosophers grapple with the meaning of self-awareness, knowing, and the soul. And somewhere in between these bounds, cognitive scientists attempt to mimic brain functions with neural nets and computer algo-

rithms. But all experts would agree that the first and most basic information on the brain and its functions must come from neurobiologists,
who probe the physical structures and chemical reactions of all manner
of animal brains—the "wetware."

Dissecting brains

Dissection of the normal human brain, a squishy three-pound mass
about the size of a head of iceberg lettuce, has contributed immeasurably to basic information on the intricate architecture of its hundred
billion cells. Though the organ's detailed anatomy is extremely complex
and laden with obscure terms, four principal structures represent the
brain's "major appliances."

(1) The central *brainstem*, which connects directly to the spinal
cord, controls basic alertness and involuntary functions of the
heart and lungs. This region is sometimes called the reptile brain
because of its similarity to the entire reptilian organ.

(2) The *cerebellum* connects to the rear of the brainstem. It coordinates the movement of muscles and may play a key role in the
transfer of memories from short-term to long-term storage.

(3) The *limbic system*, a group of small cellular structures located
above the brainstem, regulates the body's chemistry and emotional state. These components, including the pituitary gland and
the hypothalamus, employ a complex interplay of electrical nerve
impulses and chemical signals (hormones) to control the most
essential mammalian activities: sleep cycles, eating, sexual desire,
and other body functions.

(4) The *cerebrum*, the largest portion of the human brain, controls the sophisticated functions of thought, memory, and communication. The lumpy, rounded structure of the cerebrum's two
symmetrical halves provide the familiar popular icon of the brain.
The cerebrum is further divided into four lobes with distinct
functions: frontal (planning and purposeful behavior—the part
damaged in Phineas Gage's accident); temporal (hearing and

memory); parietal (the body's sensory input); and occipital (vision).

Normal brains provide a baseline, but an injured or diseased brain helps immeasurably in documenting how the wetware works. Researchers learn much about the structure and function of the brain's many parts by seeing what happens when they are damaged by stroke, trauma, or localized disease. (This strategy is strikingly similar to that of molecular biologists, who scrutinize life-forms with mutant genes to deduce what healthy genes do.) The most devastating and immediate effects occur following brainstem damage. When that region is disrupted, or its links to the spinal cord are severed, the heart and lungs cannot function and death occurs swiftly. The limbic system is also a critical region that cannot sustain much damage without severe consequences. Not coincidentally, the brainstem and limbic system are the most deeply buried and protected regions of the brain.

Other portions of the brain are somewhat more resilient to injury, and their loss may cause only partial impairment of ability. Slurred or incoherent speech, blurred vision or blindness, partial loss of memory, and changes in personality or mood often follow a stroke or other trauma to the cerebellum or cerebrum. More severe damage to these regions may result in a coma, but as long as the brainstem and limbic system are functional, life can continue in an unconscious vegetative state.

Much of our understanding of the brain's memory architecture also comes from victims with damaged or diseased brains. Brain trauma may result in sudden and serious instances of memory loss that lasts from a few moments to many months, for reasons that are not well understood. While amnesia has become a clichéd plot device in novels and movies, its awful consequences are very real and destructive to its victims. In most cases the memories are still there, hidden somewhere in brain tissues, for they eventually return in many patients. Somehow the neural pathways by which those memories are retrieved have been disrupted, and are, in an unknown way, later reestablished.

Short-Term and Long-Term Memory

The models scientists create to describe the physical reality of memories, both at the scale of cells and the whole brain, should eventually explain our shared experience of the intertwined phenomena of remembering and forgetting. Everyone knows that some memories are more persistent and more readily recalled than others. Like Charles Foster "Citizen" Kane, whose dying thoughts of Rosebud inspired his final enigmatic word, most of us will carry vivid childhood scenes to our grave. Yet most of us can't remember what we had for lunch a few days ago.

To a certain extent, we can consciously choose to place some facts in short-term memory and others in long-term memory. Most memories—an old phone number or the face of a long-absent friend, for instance—will fade with disuse. Other memories are reinforced through repetition. College faculty are often discouraged to discover how proficient students become in storing information in short-term memory by cramming the night before a test. A few days after final exams they retain almost nothing—a phenomenon that argues strongly for emphasizing critical thinking skills rather than teaching esoteric facts. Memories are more strongly imprinted when connected to a hands-on activity: It's easier to remember anatomical details if you've dissected a frog.

Another curious aspect of memories is that they may change in subtle ways with time, distorting or mixing up past events and their contexts. Compare details of an event that happened several years ago with someone else who was there with you. Chances are your recollections will differ in many details. The brain can even manufacture its own false memories, or "confabulations," such as the widely publicized accounts of alien abductions and some falsely remembered incidents of child molestation. Other memories of profound traumas may be repressed, deeply buried in our psyches. Any memory model must take account of these phenomena.

Electroconvulsive shock (ECS) treatments for depression sufferers provides additional clues about the nature of short-term and long-term memory. In this extreme procedure a strong electric current causes brain seizures, which induce temporary amnesia. ECS patients lose all recollection of recent events, but long-term memories are usually unaffected. Similar tests on trained rats provide quantitative insight on this process. A rat quickly learns certain skills, such as how to avoid an electrically charged plate in its cage; one unpleasant shock is enough to form a memory that lasts for weeks. If ECS is administered a few minutes after this experience, the rat's memory is erased and it must learn to avoid the plate all over again. Wait a full hour for ECS, however, and the rat's memory has been fully implanted. Somehow, during that first hour, a memory has been formed, reinforced, and set into the hard-wiring of the brain.

Experiments with fleas, sea snails, and mice are revealing universal aspects of the long-term memory process. These animals are trained for the experiments to remember different things: odor for the flies, retraction of gills for the snails, and mazes for the mice. Still, long-term memory is associated with the same cellular processes. For example, to establish long-term but not short-term memory, genes must be turned on and proteins newly manufactured. Activation of the genes is mediated in all three animals by the same molecular regulators of gene expression. These same molecular regulators play a role that seems to be related to memory in the hippocampus. Long-term memory is strengthened when a signaling neuron is fired over and over again.

Uncovering the Pathways of Memory

The human brain is by far the most complex physical object known—with a hundred billion cells, each linked to perhaps a thousand others, vastly more intricate than the largest supercomputer. Slices of dead brain tissue establish anatomical characteristics, but they can no more reveal the mysteries of human consciousness and memories than a cadaver can divulge the desires and fears of its former owner. To understand the intimate workings of the brain—the formation of memories and thoughts—scientists must delve into the cellular structure, and even into the very molecules, that form our minds.

Permanent damage to many regions of the brain cause partial loss of memory, in keeping with Karl Lashley's experiments on rats that suggest that specific memories are not localized. The most obvious examples are cause-and-effect relationships for degenerative memory loss

such as Alzheimer's disease. The gradual degradation of neural dendrites in Alzheimer patients causes irreversible memory loss that points
directly to synapses as a key component of memory storage and recall.
What is not yet obvious, however, is whether Alzheimer's memory loss
is associated primarily with general loss of synapses throughout the
brain, or damage to one particular area.

What is now abundantly evident is that damage to a number of
small, specialized brain structures can result in severe memory impairment. A dramatic case in point is the story of a famous patient known
only as Mr. H.M. Mr. H.M. suffered from a debilitating type of epileptic seizure, a condition for which the only known cure includes removal
of a portion of the limbic system called the hippocampus, a small
curved structure that occurs in two pieces on the two sides of the brain.
In most patients only one half of the hippocampus is removed, and
normal abilities are not impaired. In the case of H.M., however, both
halves were excised.

The devastating result was that Mr. H.M. lost all ability to form new
long-term memories. Old memories from before the operation were
intact, vivid, and easily recalled. He also retained the ability to meet
new people, converse normally, and remember their names during a
short conversation. But from the moment the anesthetic wore off years
ago to this day, he has been unable to remember anything new: current
events, changes in familiar roads or buildings, new friends. Each moment for H.M. is as if being awakened from a long dream. Evidently,
the hippocampus plays a critical role in converting short-term experience into long-term memory. Our newfound knowledge about the hippocampus was won at this tragic price.

Subsequent experiments on monkeys suggest that specific areas of
the temporal lobe, area TE, are equally critical to the visual memory
associated with remembering simple objects. A normal monkey quickly
learns the rules of a game in which a peanut is hidden under new
objects placed on a table. As the number of objects increases, the monkey has to recognize which one is new to win the peanut. When area
TE is removed from both sides of the monkey's brain, it loses the ability
to distinguish old and new objects, and short-term visual memory of
simple objects is lost.

In another set of experiments, Richard Thompson of Stanford University caused rabbits to blink when a small puff of air was blown at their eyes at the same time a tone was sounded. Soon the rabbits blinked at the sound of the tone alone—a learned reflexive behavior linking hearing and the muscles of the eyelid. After a laborious process of elimination, systematically probing parts of the rabbit brain, Thompson's research team tracked down the location of this eye-blink memory. When a tiny region of the cerebellum was destroyed, the rabbit still blinked at a puff of air, but it had lost the reflexive response to the tone. Furthermore, it could not relearn the reflex. The specific ear-to-eyelid neural pathway had been permanently destroyed.

All of this research reveals that memories are real structures, imprinted on the brain. Specific pathways linking portions of the hippocampus, cerebellum, and cerebrum are evidently responsible for storing and retrieving memories. It's even possible to document specific crossroads in a few of these complex neural pathways. Yet, in spite of such suggestive findings, the architecture of memory is almost completely unknown and remains one of the central unanswered questions of neurobiology.

The Dynamic Brain in Action

Memory research will continue to be fueled by studies of animals with ever more precisely excised bundles of neurons and humans with damaged and diseased brains. Much of what we know about memory as a physical process comes from these pioneering studies. But, in a sense, these experiments are something like trying to learn how a 747 flies by sawing off chunks of metal and snipping wires to see if it will crash. We cannot hope to reveal intimate details of how the brain works just by selective destruction.

Though still in their infancy, a promising array of sensing techniques now allows researchers to examine the dynamic brain in action. Detailed anatomical structures of the living brain are routinely documented by complementary methods, including computerized tomography (CT scanning) and magnetic resonance imaging (MRI). Computerized tomography employs numerous different X-ray pathways to construct a detailed three-dimensional map based on the density of matter inside the skull. The MRI accomplishes the same feat without radiation by placing the brain in a strong magnetic field that causes specific molecules (usually water) to vibrate or resonate, which in turn alters the magnetic field. By measuring magnetic field variations, the MRI scan creates a map of water distribution—another indicator of tissue density.

Researchers can also turn to their advanced experimental hardware to study the brain in action. MRI, for example, can be used to map the distribution of metabolic activity during thought. In a recent series of experiments by Avi Karni and colleagues at the National Institute of Mental Health in Bethesda, Maryland, researchers monitored brain activity in subjects as they practiced touching fingers to thumb in an exact sequence. After weeks of daily practice, subjects had improved their speed and accuracy, and a much larger region of the brain's primary motor cortex that regulates voluntary motion was engaged during the process. Karni concluded that long-term learning alters brain pathways.

Another promising avenue of brain remote sensing is the extraordinary antimatter technology of positron emission tomography, or PET scanning. PET scans take advantage of the close relationship between cellular activity and blood flow. Volunteers receive an injection of radioactive glucose, which emits positron radiation (positrons are the antiparticle of the electron). Glucose is fuel for brain cells, and it quickly concentrates anywhere that brain cells are active. By measuring the emission of positrons from different regions of the brain, researchers can identify the locations of the most intense brain activity. PET scans reveal, for example, that men and women use different parts of their brains when crafting rhymes, performing math problems, judging emotions, or just thinking about nothing in particular.

Ultimately, what we really want to know are the exact neural pathways by which lightning-fast electrical impulses travel. While no device is close to providing that kind of resolution in space and time, experimental technology is improving at a rapid pace. The most familiar monitoring technique, electroencephalography, or EEG, has been around since the late 1920s, when Austrian psychiatrist Hans Burger first measured electric fields emanating as a result of neural activity in the brain. Researchers discovered that a variety of brain abnormalities cause distinctive EEG patterns, and the device soon became a routine diagnostic tool.

The EEG is completely passive, requiring only that an array of electrodes be attached to the patient's head. One of the great advantages of the EEG is that it produces a continuous dynamic measure of brain activity on the scale of thousandths of a second—time intervals typical of individual synapse events. Characteristic brain patterns can be monitored during various activities such as reading, calculating, or sleeping, and for specific stimuli such as a pinprick or a flash of light. The method has traditionally been limited, because no one specific response can be fully isolated from the constant jumbled background of brain "noise." But new protocols that employ a hundred or more precisely positioned electrodes and sophisticated computer analysis may lead to a detailed picture of signal pathways in the brain.

Magnetoencephalography, or MEG, operates on a principle like EEG, but monitors variations in the brain's magnetic field rather than electric field. The main advantage of MEG over EEG is its ability to pinpoint the location of brain activity—a task of critical importance in finding the source of brain short-circuiting in epileptics. A major drawback has been the device's expense and technical complexity, but new superconducting magnetic sensors may enhance the capabilities and decrease the cost of MEG research.

BEYOND MEMORIES: BEHAVIOR AND CONSCIOUSNESS

Memory studies represent only a small fraction of the work that neurobiologists do. Ongoing studies of human behavior and abilities address endlessly fascinating questions about the brain's control of who we are. Here are a few examples.

The Two-sided Brain

Some of the most remarkable insights about the brain and its functioning have come from studies of patients with damage to the connecting neurons between the left and right brain. Upon close inspection, the two halves of the brain appear as mirror images of each other. They are essentially the same size, the same weight, and they have similar patterns of bumps and convolutions. Most anatomists of the early nineteenth century assumed mirror symmetry in the brain's functions as well, but this conventional wisdom succumbed to a growing body of anecdotal evidence from stroke victims and other brain-damaged patients. The celebrated French neurosurgeon Pierre Paul Broca and his mid-nineteenth-century contemporaries noted a curious trend: Damage to the left hemisphere often results in speech impairment, while right-hemisphere damage is more likely to cause attention deficit and difficulties in retaining spatial information. Physicians were forced to conclude that the two sides of the brain play dramatically different roles.

The most detailed and compelling evidence for the distinct roles of the brain's two hemispheres has been provided during the past half century by studies of split-brain patients. Research in the early 1940s revealed that severe epileptic seizures sometimes result from intense unregulated electrical impulses passing between the two hemispheres through a connecting bundle of nerves called the corpus callosum. Physicians had already observed that damage to the corpus callosum reduced the incidence of epileptic attacks in some patients. Thus, in an effort to alleviate the crippling symptoms of severe epilepsy in which victims suffer numerous seizures every day, doctors resorted to severing all connections between the halves of the brain, including the entire corpus callosum as well as other smaller connecting bundles of nerve tissue. With this drastic procedure the epilepsy was usually cured, and while relatively few dramatic changes in everyday behavior and abilities were noted, new curious phenomena about the two sides of the brain were revealed.

The two halves of a split brain can be studied independently because of an intriguing anatomical quirk. All sensations to the left side of the body, including all visual inputs to the left halves of both eyes, are channeled exclusively to the right brain, and vice versa. In split-brain patients the right hand quite literally may not know what the left hand is doing. Neurobiologist Roger Sperry, who won a Nobel prize for his innovative split-brain research, has concluded that these patients develop two independent "realms of consciousness," one for each side of the cerebrum.

In their classic text *Left Brain, Right Brain,* Sally Springer and Georg Deutsch describe a typical experiment in which a split-brain woman is asked to focus on a small black dot in the center of a white screen. An image of a cup is briefly flashed to the right of the dot, sending visual information exclusively to the left brain. The woman easily uses words to describe the object because the left brain has no trouble using language to associate words with the image. When a picture of a spoon is briefly flashed to the left of the dot, however, the woman says that she sees nothing. Nevertheless, she is able to reach under the screen with her left hand and correctly select the spoon from several different items. While the right brain lacks the verbal skills necessary to describe the utensil, it excels in solving spatial problems. When asked to pick out the same object with her right hand (i.e., left brain), she is unable to do so.

In another test, a picture of a nude woman is flashed to the left of the dot. The woman claims to have seen nothing, but she blushes, begins to giggle, and exclaims: "Oh, doctor, you have some machine!" The right brain processed the racy image sufficiently to evoke an emotional response, but not a verbal one. These and many other creative experiments are gradually painting contrasting pictures of the left and right brain.

Occasionally, a split-brain patient will find his or her two halves in conflict. In one bizarre case, a man became furious at his wife and attempted to hit her with his left hand, while restraining the aggressive arm with his right. A female patient reported similar conflicts while getting dressed; each hand selected a different outfit from the closet, and neither would let go. Such incidents provide anecdotal evidence that communication between the hemispheres plays a key role in resolving inner conflicts. Each hemisphere moderates the actions of the other.

In spite of the dramatic discoveries associated with research on split-brain patients, many questions about brain asymmetry remain unanswered and intense research efforts are advancing this knowledge on several fronts. Neurobiologists still don't know why some people are right-handed and others left-handed, or why lefties are somewhat more likely to become artists—and, strangely, much more likely to suffer from immune system disorders. They have no idea how the brain's two sides cooperate in creative thought, nor do they yet know the relative roles of the hemispheres in music, memory, and emotion. And, in one of the most surprising puzzles raised by split-brain research, scientists are still unsure to what extent, and why, brain asymmetry differs in men and women.

Do Men and Women Have Different Brains?

Everyone knows that men and women behave differently (even if they are reluctant to admit it on ideological grounds), but *why* they are different presents one of the knottiest problems in science. Some observers argue that stereotypical gender differences in intuition, body language, favorite toys, and emotion are fundamentally a matter of environment and socialization, while other scholars lean toward a more deterministic biological explanation. Research biologists, sociologists, and psychologists, who invariably come to this question with a conflicting mix of open-minded curiosity and deeply ingrained preconceptions, hope studies of the brain will provide some unambiguous answers to this contentious issue.

Broad-based testing of men and women reveals an intriguing trend: Gender differences in abilities closely match the distinct characteristics of the brain's two hemispheres. Women tend to be more proficient in left-brain abilities. They are on average superior in language skills, including speech, grammar, and fluency, as well as at tasks requiring fine motor skills, arithmetic calculations, and matching of shapes. Standardized tests also indicate that men are better on average in right-brain-dominated activities such as mechanical skills, spatial problems, and abstract reasoning.

In spite of such trends, until recently relatively few studies addressed the possibility that the brain physiology of men and women differs. The first suggestive evidence came from research on differences between half-brain-injured men and women. Men usually display a pattern of lost spatial perception following right-brain damage, and lost verbal ability following left-brain trauma. Women, however, typically suffer much less loss of ability, regardless of the side of brain damage. The implication is that right brain and left brain functions in men are largely separate, while in women considerable overlap occurs.

These differences are mirrored by tantalizing discoveries on gender differences in brain anatomy, though such studies are few and far between. A 1991 study at UCLA confirms sketchy nineteenth-century reports that women's brains have on average a thicker corpus callosum connecting the cerebrum's two hemispheres, suggesting greater connectivity. Much more basic anatomical work comparing details of men's and women's brains is needed; periodic headlines announcing important new findings can be expected.

While much of this research is intriguing and suggestive, all questions about gender characteristics and the human brain face a serious experimental caveat. Brains, like people, are not standardized: Each one is different, with a unique combination of size, shape, abilities, and limitations. Some women are better at spatial thinking and sports than many men, while some men are more emotional and verbally skilled than many women. Human characteristics are fuzzy, not subject to simple quantitative measurements. While it would be absurd to deny the existence of behavioral differences *on average* between men and women, it is difficult to quantify those differences, much less point to unambiguous cause-and-effect relationships.

What Is Consciousness?

Whatever the technologies, and whatever clever experiments are devised to exploit them, a hundred years from now scientists will still be trying to understand the workings of the human brain. Perhaps by then the question "What is memory?" will have been answered in some detail. But it seems likely that deeper mysteries associated with consciousness will remain to take its place.

The problem of consciousness has been pondered by myriad scientists and philosophers, from avowed reductionists who expect that thought and emotion can be explained by neurons alone, to skeptics who deny any hope of physical understanding. University of California philosopher David Chalmers adopts a useful intermediate view by dividing the question "What is consciousness?" into what he calls the "easy problem" and the "hard problem."

The easy problem focuses on mechanics of consciousness: How can humans isolate external stimuli and react to them? How does the brain process information to control behavior? How can we articulate information about our internal state? Neurobiologists have long tackled aspects of these questions, which are amenable to systematic study in much the same way that researchers probe the physical mechanisms of memory. Perhaps, with many decades of intense research, such questions can be answered.

The hard problem, on the other hand, relates to the intangible connections between the physical brain and self-awareness, emotion, perception, and reasoning. How can music evoke a sense of longing, or a poem deep sadness? How does reading a book stimulate curiosity or frustration? What are the physical structures and processes that produce love, fear, melancholy, or greed?

A now-classic thought experiment highlights the difference between the easy and hard problem. Imagine a researcher who is the world's expert on color vision, but who has spent her entire life in a black and white room. She has learned everything possible about how the brain receives visual information, processes the images, and articulates what it perceives. She has completely solved the easy problem for color vision. But in her black-and-white world, no amount of study, no complex equation or descriptive prose, can reveal to her what it is like to experience the color red. Chalmers concludes: "There are facts about conscious experience that cannot be deduced from physical facts about the functioning of the brain." The hard problem, for the time being, is beyond the realm of science.

Some researchers believe that in due time an understanding of consciousness will follow naturally from research on the physical brain. Bold and speculative books by Francis Crick, Gerald Edelman, and Roger Penrose, among others, offer ideas on how such a synthesis might be achieved with current methodologies. Others argue for a radically new perspective. Chalmers, for example, makes the startling proposition that consciousness must be accepted as a characteristic of the universe *completely distinct* from previously recognized physical attributes, such as matter, energy, forces, and motions. Perhaps, he says, consciousness is an (as yet unrecognized) intrinsic property of information.

What is consciousness? Scholars cannot yet even agree on what exactly the question means, much less imagine the form an answer might take. For as far into the future as anyone cares to foresee, this greatest mystery of the human mind may remain.

BEHAVIORAL GENETICS: ARE THERE SOME QUESTIONS THAT SCIENTISTS SHOULDN'T ASK?

> Perhaps most disturbing to our sense of being free individuals,
> capable to a large degree of shaping our character and minds, is
> the idea that our behavior, mental abilities, and mental health can
> be determined or destroyed by a segment of DNA.
> —Torsten N. Wiesel, *Science* (1994)

The Crime Conference

STOP NEO-NAZI PSEUDOSCIENCE! the placard proclaimed as protestors shouted, "Maryland conference, you can't hide/We know you're pushing genocide!"

The much-delayed and vehemently criticized conference "The Meaning and Significance of Research on Genetics and Crime" finally took place in September 1995 in a confrontational atmosphere atypical of scholarly gatherings. For half of the conference's seventy participants it was a chance to outline a growing body of data that suggests links exist between heredity and behavior. For the other half, invited to balance the controversial agenda, the meeting provided a forum to denounce behavioral genetics, a field its most outspoken critics describe as based on shoddy experimental evidence and driven by a reprehensible ideology.

Proponents of behavioral research plead that heredity plays a central role in many kinds of human behavior, and that such research is essential if we are to solve many of society's pressing concerns. If genetic markers for tendencies to antisocial behavior can be identified at birth, they suggest, then steps can be taken at an early age to alleviate environmental factors that promote these tendencies. Many critics counter on experimental grounds, claiming that behavioral traits are impossible to quantify in any meaningful way, that environmental factors are far too complex to identify, and that the human genome encodes characteristics that are much too adaptable to allow for simple correlations. Others argue that because most arrests and convictions for violent crimes occur among minority groups, any attempts to identify links between heredity and crime are inherently racist and should be stopped.

In such a contentious atmosphere, it is surprising that one area of common ground has emerged. While everyone agrees that behavioral scientists must be vigilant to monitor how such studies are used, they also accept that it is not reasonable to ask scientists to ignore a fundamental unanswered question.

THE LIMITS OF
ETHICAL SCIENCE

Forbidden Knowledge

Many scholars claim that knowledge in and of itself is neither good nor evil. Indeed, some thinkers argue that a democracy can thrive only when all kinds of knowledge are available freely and openly to all citizens. Any suppression of knowledge, and by extension any proscription on questioning, erodes the foundation of our most cherished political ideals. On the other hand, all knowledge is discovered and used within the context of a society. The questions scientists choose to ask, the methods by which they answer those questions, and the uses to which this knowledge is applied must be subject to unflagging ethical scrutiny. Scientists cannot expect to pursue their research divorced from the society that supports the effort, and they must live with the consequences of the expanding horizon of knowledge.

This concept is not new. Thinkers from Socrates to Salman Rushdie have been punished for addressing questions deemed to be forbidden territory by particular, vocal elements of society. Medieval studies of human anatomy drew long and harsh attacks from conservative theologians. Galileo Galilei was jailed and threatened with torture for reporting his telescopic observations of the heavens. Giordano Bruno was burned at the stake in 1600 for proposing the existence of extraterrestrial life. Charles Darwin was attacked for his atheistic theory that evolution is driven by an undirected violent struggle for survival, and Soviet biologists were persecuted for any sympathetic reference to genetic theories.

Societies change, and so do their restrictions on knowledge. Today the physical sciences are not taboo, and most people calmly accept the basic tenets of Galileo and Darwin. Startling discoveries of physics, cosmology, and earth science are viewed almost universally in a positive light, and, in spite of the possible dangers of some new chemicals, the public is generally supportive of research on novel materials. But even in our enlightened age, some people point to research topics that scientists should not pursue, and questions they should not ask.

Science is a two-edged sword. On the one hand, science and technology contribute directly to economic well-being, national security, and longer, more productive lives. Science teaches us how to cure disease, to feed the hungry, to live richer and more productive lives, and to discover the magnificent order underlying life and the physical universe. But scientific knowledge can also expose the world to new hazards, such as chemical pollution, radiation, and nuclear weapons. The tragic consequences of knowledge unchecked appear as a recurrent moral lesson in literature and film, from the horrific punishment of Prometheus for granting humans the knowledge of fire, to the gory demise of Jurassic Park's genetic engineers. Countless fictional scientists have brought disaster upon themselves and those around them by probing "forbidden" subjects. These obsessed researchers are typically portrayed as oblivious of the devastating consequences of their uncritical pursuit for knowledge. While such facile stereotypes bear little resemblance to reality, it is worthwhile considering the difficult, even intractable dilemmas that scientific knowledge can pose for society.

The paradoxical view of the scientific enterprise—that "In much wisdom is much grief: and that which increaseth knowledge increaseth sorrow" (Ecclesiastes 1:18)—arises in part from a suspicion that scientists, while extremely effective at arriving at answers, may, at times, be irresponsible when it comes to asking questions. Though professional researchers tend to rankle at the suggestion, many nonscientists suspect at a gut level that there are some questions it is much better not to ask.

Setting Limits

Many contemporary criticisms of research focus not on the questions being asked, but rather on the means by which legitimate questions are being answered. Our society places strict prohibitions against involuntary or potentially dangerous experimentation on human subjects. Research on biological and chemical weapons is justifiably banned by international treaty, as are many forms of study on human subjects.

The ethical dilemma facing animal rights activists and animal re-searchers—the extent to which human rights apply to animals—is less straightforward. Many of the most visible and effective protests against members of the contemporary scientific research community have been organized by the diverse animal rights movement. Any attempt at guidelines is complicated by the wide variety of animals used, the ex-tent of pain and suffering to which they are subjected, and the urgency of the research. Few people, even outspoken animal rights workers, ob-ject to research on roundworms and fruit flies, even if studies result in the mutation or death of the animals. Research on monkeys or dogs, however, often raises serious objections even if the animals are healthy and well treated. Few researchers would hesitate to study the behavior of cockroaches in a maze, but no reputable scientist would attempt to justify neurological studies that require subjecting chimpanzees to in-tense pain for prolonged time periods.

Genetic engineering, by which the chemical mechanisms of organ-isms (including people) are altered, raises similar troubling ethical is-sues. While humans have been manipulating genes for thousands of years in selective breeding efforts, the ability to target and alter specific genes at will raises two broad families of thoughtful reservations in the minds of many citizens. The first and most basic misgivings relate to the ethics and potential dangers of producing genetically modified or-ganisms. In a recurrent theme, echoing Mary Shelley's cautionary tale of Dr. Victor Frankenstein and his monstrous creation, modern authors exploit fears of genetically modified organisms, ranging from virulent bacteria to savage dinosaurs. And while Michael Crichton's vision of a world run amok with resurrected carnivores is little more than a fanciful parable, the threat of genetically engineered biological weapons is both real and scary.

The promise of genetic engineering to solve some of humanity's most pressing crises of disease and hunger complicates the ethical choices and negates calls for an absolute ban on such research. Most people when offered the hope of a miracle cure for a dying child will not ask by what path the end is achieved. As in all other medical research, informed oversight and prudent applications are deemed suffi-cient safeguards for the time being. But a society that accepts with widespread, ringing enthusiasm advances in gene therapy for medical purposes must understand that the more controversial applications to genetically engineered strains of *E. coli* and juicy varieties of tomatoes are based on the identical technology. What we choose to accept as worthwhile and ethical is a question of degree rather than of absolutes.

WHERE GENETICS
AND BEHAVIOR MEET

Ethical objections to animal research and genetic engineering are based primarily on experimental methods. For the most part, the underlying questions and their motives—curing disease, feeding the world, and understanding the processes of life—are worthy and not under attack. But one long-standing scientific topic is, in the minds of some, as close to a forbidden question as anything modern researchers can pursue. Dare scientists ask to what extent human behavior is dictated by heredity?

One fact must be made clear from the outset: There exist incontrovertible links between human behavior and genes. These links were discovered essentially by accident, in studies of drugs, illness, and other seemingly nonbehavioral phenomena. A defective gene that blocks hunger suppression may lead to overeating. Children with Down syndrome, a genetic disease, are exceptionally kind and gentle, while those with Tourette's syndrome may display extreme antisocial behavior. And hereditary forms of blindness, muscle disease, and brain disorders also have obvious behavioral consequences. It is also worth noting that none of these genetic conditions was originally studied by behavioral scientists—links between heredity and behavior were recognized only later. That such links exist is not a matter of much debate.

Equally important is the recognition that the familiar concept of "one gene, one disorder"—that defects in single genes are necessary and sufficient to cause specific diseases—cannot be extended to most behavioral traits. While specific errors on specific genes cause Huntington's disease, sickle-cell anemia, cystic fibrosis, and many other diseases, there is not a single "happy gene," "music gene," or "crime gene." Rather, complex behavior and ability must arise from interactions among the products of a collection of several genes, known as quantitative trait loci, or QTL in the jargon of modern behavior research. Links between heredity and behavior must address a QTL that also responds to many environmental factors.

The Legacy of Eugenics

There was a time when eugenics, the science of understanding and improving hereditary qualities, especially those of humans, was a highly respected field. Indeed, during the 1910s and 1920s, many political and social leaders viewed eugenics as society's best hope for helping each individual to achieve his or her potential. Our own institution played a central role in the movement, housing the Eugenics Record Office in its Department of Genetics and sponsoring global conferences on eugenic research. Carnegie Institution President John C. Merriam stated in 1921: "There is no group of questions more significant in the complicated organization of human society than those concerning the meaning and the possibility of direction and control of inheritance in man." By identifying and correcting genetic predispositions, the argument went, each individual's education, career, and personal choices could be customized to maximize opportunities for a healthy and happy life.

How quickly this flawed and naive vision became perverted. The United States restricted immigration of presumed low IQ groups and instituted wholesale segregation based on supposed links between genetics and behavior. Twenty-four states in the U.S. employed intelligence tests to justify enforced sterilization. And then the Nazis began their ultimate crime of eugenic terrorism, rationalized on the basis of "scientific" discoveries. The implications of the eugenics question have been forever changed.

In the context of such a vile history, suggestions of a link between genetics and any aspect of behavior or ability are understandably met with deep reservations. Critics wonder why these questions are being asked, and to what use the information will be put. For two decades after World War II, studies of heredity and behavior were all but abandoned, as behavioral scientists espoused the philosophy of "environmentalism"—that environment accounts for virtually all behavioral traits. Schizophrenia and learning disorders, for example, were blamed on inadequate parenting and inappropriate schooling. The pendulum began swinging back toward a balance between nature and nurture in the 1960s and 1970s, in large measure due to psychiatric research into the behavior patterns of biological versus adopted children. The importance of the body's chemical balance was also underscored by a growing awareness and use of behavior-altering drugs.

Today, a new generation of genetic researchers are tackling the old questions—searching for genetic influences on human behavior and abilities—with new methodologies and new rhetoric. However, virtually all this research is carried out with the understanding that genes and environmental influences interact to influence behavior. It is no longer a question of nature or nurture; it's nature *and* nurture.

Animal Behavioral Genetics

Animals provide an ideal means to study behavior without raising many of the difficult social and political concerns associated with human behavioral research. Centuries of observations in selective breeding of domestic animals demonstrate beyond any reasonable doubt that behavior is determined in part by heredity. Selective cross-breeding has led to gentleness in domestic dogs and aggression in killer bees. Animal behaviorists now employ the sophisticated techniques of molecular genetics to identify specific genes and their roles. Many of these researchers, furthermore, believe that understanding how genes affect mating rituals in fruit flies, alcoholism in rats, or aggression in mice may eventually provide important clues to origins of human behavior.

The research strategies are conceptually simple, even if the experimental details are not. In pioneering studies begun in the 1960s, Seymour Benzer of Caltech zapped fruit flies with radiation and watched for mutant offspring exhibiting unusual behavior. While most flies were unaffected, a few would fly, feed, rest, or mate in odd ways. In three decades of research, observing and tracking down such aberrant behavior, Benzer and others have identified numerous genes that participate in the routine functioning of the fly's nervous system.

A similar strategy has been applied by many biologists to nematodes (roundworms), which are distinguished by their exceptionally simple nervous systems. With only 302 neurons and about 5000 synapses, the neural net of the roundworm *Caenorhabditis elegans* has been completely mapped. In some experiments, behavior-altering mutations are induced, and those changes are correlated with aberrant genetic sequences. More than 250 genes that affect behavior have been spotted so far, and the list is growing. Other research employs a laser microbeam that targets and kills specific individual neurons, thus impairing behavior through physical trauma. Ultimately, by documenting every aspect of this simple living system, biologists hope to unravel the complex interactions of physiology, heredity, and environment.

Mammals display behaviors far more complex and less amenable to comprehensive study than roundworms, but mice and monkeys are more likely to shed light on hereditary influences in humans. In one widely used experimental protocol, biologists cross-breed two pure strains of animals that display contrasting behavior—pure-bred highly aggressive mice crossed with docile mice, for example. Several generations later, the offspring, which have complexly recombined genes of the original two strains, will display a range of aggression. Researchers rank each individual mouse according to its aggressiveness, and then analyze for diagnostic DNA sequences that indicate the presence of specific genes. By searching for correlations between these DNA markers and behavior, a hereditary component of behavior can be tied to specific genes.

A team of British and American scientists led by behavioral geneticist Jonathan Flint has employed this protocol to identify DNA segments that are found primarily in unusually fearful mice. Laboratory mice display a wide range of behavior when placed in a brightly lit, open area. Some individuals actively explore their new surroundings, while others cower in one corner, "defecating copiously." Flint and coworkers found a high degree of correlation between fearfulness and three specific DNA segments. They conclude that these genetic characteristics "may be related to human susceptibility to anxiety or neuroticism."

Studies of "knockout mice," in which researchers inactivate a specific gene, also reveal clear links between mouse behavior and genes. In 1993 a team led by French microbiologist René Hen targeted a gene that activates specific neuroreceptors for the chemical serotonin, which transmits signals in both mice and human brains. These knockout mice displayed extreme aggression, leading the French scientists to speculate that similar defects might contribute to abnormal human behavior.

In all of these animal experiments, whether on fruit flies, round-worms, or mice, the exact genetic makeup of the subjects can be determined and their environments can be precisely controlled from birth. Such control is not possible with human subjects.

Assumptions

Human subjects are vastly more difficult to study than mice. For a time, perhaps a few decades, this complexity will grant society some ethical breathing room. Any controversial conclusion about heredity and human behavior will be subject to scathing attack on the grounds of questionable experimental methods, such as sample size and lack of sufficiently rigorous control. Those who wish to accept or ignore the conclusions of such studies for ideological reasons may, for a time, rationalize their choice on "scientific" grounds. (The recent debate over genetic links to homosexuality, based on molecular genetics research by Dean Hamer of the National Cancer Institute, provides a widely publicized and telling case in point.)

But society cannot dodge the moral dilemmas posed by behavioral genetics forever. Like it or not, we may soon be able to answer such questions in a quantitatively rigorous way. The three critical assumptions underlying behavioral genetics are (1) that behavioral and personality traits can be ranked, at least in a qualitative way; (2) that crucial environmental factors influencing behavior can be identified; and (3) that distinctive genetic characteristics of individuals can be quantified.

1. Ranking: The assumption that traits can be ranked is in accord with everyday experience. No absolute numerical scale is likely to emerge, but everyone knows individuals with varying abilities in music, mathematics, athletics, puzzle solving, and so forth. One person is more outgoing, organized, and good-natured, while another is more reserved, impulsive, and quarrelsome.

Researchers in behavioral genetics, who claim that a variety of observations and tests are sufficient to establish these and other aspects of behavior, generally agree that personality rests on five broad traits: extraversion (outgoing, decisive, and persuasive), neuroticism (emotional, irritable, and anxious), conscientiousness (organized, practical, and dependable), agreeableness (good-natured, warm, and kind), and openness (curious, insightful, and imaginative). To one extent or another, each of us displays a combination of these five traits, and the assumption is made that a qualitative ranking of individuals can be achieved. Extraversion, which encompasses a range of behavior from cringing fearfulness to self-confidence to violent aggression (and thus may relate closely to antisocial criminal behavior), has received the most attention from researchers in recent years.

A similar strategy has been applied to recent studies of intelligence. While a single number representing IQ has been shown to have little meaning, attempts to quantify several different aspects of cognitive ability—memory, spatial reasoning, and verbal ability, for example—have gained a receptive hearing. Behavioral researchers admit that any effort to rank behavioral traits incorporates unsettling subjective aspects, but they argue that common experience demonstrates that individuals do possess such traits to varying degrees.

2. Environment: The second assumption underlying behavioral genetics, that important environmental factors influencing behavior can be identified, is a matter of considerable debate. Many factors, including nutrition, exposure to drugs, sleep patterns, and stress, are well known. Others influences such as visual stimuli in infancy, pollutants in air and water, exposure to sunlight, and types of physical activity may be much more subtle and difficult to track down. The National Cancer Institute's Dean Hamer claims that "environment" includes such disparate factors as "the flux of hormones during development, whether you were lying on your left or right side in the womb, and a whole parade of other things." He concludes that "genes might have a somewhat different effect on someone in Salt Lake City than if that person were growing up in New York City."

3. Genes: The third assumption of behavioral genetics, the conviction that we can document genetic characteristics of an individual, until recently was based more on anecdote than reproducible scientific observations. Similarities within and between family members, particularly siblings, have provided most of the data, but the only possible rigorous control subjects for genetic research are identical twins with identical genomes. Studies on the behavior of identical twins reared apart versus those reared together have offered the best hope for quantifying the nature versus nurture debate. For a time the widely publicized fraudulent research of British behavioral scientist Cyril Burt, who invented twins to suit his preconceptions, tarnished the public perception of such efforts. Recent investigations, notably by the University of Minnesota's Center for Twin and Adoption Research, are expanding the twin database and once again legitimizing this classical behavioral research.

Soon the rapidly improving ability to locate and sequence genes may provide the ultimate validation of this third assumption. Eventually, we will be able to determine each individual's exact, unique genome—all 80,000 genetic words. The day will come when an entire human genome can be sequenced automatically, perhaps in a few days. Rather than the single human genome sequence envisioned by the Human Genome Project, we will have access to tens of thousands of individual genomes, each a detailed chemical map of a unique individual.

If all three assumptions of behavioral genetics are realized, and there is every reason to believe that they might be, then statistical tests may establish definitive links between heredity and personal traits. Databases containing thousands of individual genomes will be correlated with distinctive traits—musical ability, aggressiveness, mathematical skills, or sexual preference, for example. Given a sufficient number of individual genomes and carefully documented traits for each of those individuals, unambiguous and statistically rigorous correlations might emerge.

Scientists, who come to these questions with their own ideological agendas, are deeply divided over the accumulating evidence of behavioral genetics. At one extreme, our genetic heritage may be viewed as a tool kit, a package of remarkable adaptability that equips us from birth to cope with the unpredictable range of opportunities and vicissitudes that confronts us throughout life. According to this view, each person's behavior and characteristics are determined by complex interactions between environment and heredity—genes turn on and off as required, allowing us to change our behavior as circumstance dictates. Most behavioral researchers interpret the data in this light; some go so far as to claim that we are who we are because of an equal, inextricable mix of heredity and environment.

Other investigators interpret the data differently, and argue that our destinies are shaped largely by genetic factors, both good and bad, that preordain whom we will become from the moment of conception. They endorse the concept of "genetic essentialism"—that attitude, abilities, emotions, and desires are inextricably linked to a nucleotide sequence, and that environmental factors play a relatively minor role. They conclude that people are destined to be happy or sad, curious or dull, violent or passive, irrespective of their environment. From this perspective, even our adaptive responses to environmental changes are preprogrammed in our genes. While these scientists do not expect to find a single "music gene" or "crime gene," they suspect that a concatenation of many genetic characteristics might be highly correlated with abilities and behavioral characteristics. Perhaps a certain combination of distinctive genes on chromosomes 3, 7, 17, and 21 will be found to appear in most great composers. Perhaps genes on chromosomes 5, 11, and 18 are common in individuals prone to worry.

The Dilemma of Behavioral Research

Scientists, like the public at large, are strongly divided over the ethical implications of behavioral genetics. Proponents see this research as the best hope for understanding and ameliorating antisocial behavior. They point to the complex interactions of genes and environments—many genetic traits are expressed only under specific conditions, for example under stress or malnutrition. By learning the genetic *and* environmental basis of addiction or violence, these experts insist, society will be in a much better position to cope with these problems. Behavioral researcher Xandra O. Breakfield, who studies links between heredity and violence as part of a project at Massachusetts General Hospital, claims her work is designed "not to push people into trouble but to pull them out."

Other scholars passionately disagree. Peter Breggin, the outspoken director of the Center for the Study of Psychiatry in Bethesda, Maryland, argues that "behavioral genetics is the same old stuff in new clothes. It's another way for a violent, racist society to say people's problems are their own fault, because they carry 'bad' genes." For Breggin and many other concerned observers, the question is not so much whether heredity and behavior are linked, but rather the chilling prospect of scientific validation for new forms of genetic discrimination. (The hidden agenda of some behavioral studies is vigorously attacked in *The Mismeasure of Man*, Stephen Jay Gould's disturbing exposé of prejudice and fraud in the history of intelligence research.)

Researchers, and the society that supports them, must together examine the potentially disturbing consequences of behavioral genetics. Assume for the moment that a wide range of positive correlations is found between genetic markers and the predisposition for specific genetic diseases, mental abilities, physical traits, and behavioral characteristics. Also assume that a fetus can be tested quickly and reliably for these genetic markers. (We emphasize that scientists are only just beginning to possess such capabilities for a few specific genetic diseases, but the potential for more sweeping advances is real.) Where does this research lead?

The case of genetic diseases is least problematic. Upon learning that a fetus carries a serious defect, parents may eventually have the option of an appropriate therapy to treat, or perhaps even to correct, the problem. If no cure is possible, and the disease is sufficiently serious, some parents may elect to terminate the pregnancy. In any event, foreknowledge of such genetic disorders will allow families and their physicians to make informed choices. For these reasons, research on identifying and curing genetic diseases will continue to be a major focus of modern medical research (see Chapter 9).

Identification of extraordinary abilities is more problematic, though hardly a source of great moral angst. Most parents do not object to being told that their child is a potential musical prodigy or gifted athlete. Schools have a long tradition of administering diagnostic tests in preschool or elementary school. Society must decide whether tests administered *before* birth differ in any fundamental way, and, as now, it must decide how that information will be used. We would argue that while it is worthwhile to provide gifted children with opportunities to develop those talents, every child should be given the chance to succeed or fail in many pursuits. Many mediocre singers find music their greatest joy in life, while some brilliant symphonic musicians despise the demands of their profession.

Visions of a future in which comprehensive genetic evaluation of infants is routine leads to unsettling scenarios. When asked to make a prediction about scientific advances of the next century, Harvey F. Lodish of the Whitehead Institute for Biomedical Research in Cambridge, Massachusetts, imagined a time when an embryo's complete DNA sequence could be analyzed and understood. His extreme prediction: "All of this information will be transferred to a supercomputer, together with information about the environment—including likely nutrition, environmental toxins, sunlight, and so forth. The output will be a color movie in which the embryo develops into a fetus, is born, and then grows into an adult, explicitly depicting body size and shape, and hair, skin, and eye color. Eventually the DNA sequence base will be expanded to cover genes important for traits such as speech and musical ability; the mother will be able to hear the embryo—as an adult—speak or sing." Though this technology is at best a distant dream, most parents would reject such a premonition, for it would rob us of a portion of the expectation and uncertainty, the wonder and joy that comes with the unfolding of a unique human being. (Of course, twenty years ago the idea of knowing the sex of a baby before birth was equally unthinkable, and now such information is routinely provided.)

Finally, and most troubling in the continuum of genetic predictions, are traits society deems undesirable. How should society deal with children shown to have a genetic predisposition for substance abuse, delinquency, or violent behavior—if such a correlation is found? When it is possible to know and interpret every child's genome, what use do we make of our knowledge of a lurking potential for self-destructive behavior? If history is any indication, society will not ignore such an awesome capability.

Those who now advocate social change—reforms in education, improvements in nutrition, and reductions in poverty—will find ample ammunition in any correlation between genes and behavior. If genetic traits influence behavior, then changing a child's environment represents our only realistic hope of ameliorating these predispositions. If, on the other hand, one adopts a more conservative agenda, then findings of behavioral genetics will be equally reassuring. Genetic traits, it will be argued, absolve society of responsibility for poverty, crime, and drug abuse.

Eventually, as techniques of genetic engineering become more effective, some social engineers will advocate society's right to fix whatever we deem to be a genetic defect or deficiency. Pop in a new base pair here and there and, *voilà*, this or that disease is cured, a few higher points on the IQ test, and an "improved" personality to boot. Someday we may have a gene therapy for violent criminals—a nose spray, perhaps. Some fine day society may have reached such a state of enlightenment that we will be prepared to deal with these contentious issues. But we are not there yet.

All viewpoints in the behavioral genetics debate have some validity. Social activists are correct that there are grave dangers inherent in any research that provides a scientific basis for prejudging an individual's abilities and personality. Society must remain vigilant against any misuse of such knowledge. Skeptical scientists are correct that present methodologies at times lack rigor and leave much to be desired. No one has yet come close to understanding the complex interplay between heredity and environment for any one human trait. And behavioral geneticists themselves are justified in their continued pursuit of a compelling unanswered question, even if experimental techniques are not yet able to provide definitive answers, and even if society is not yet prepared to deal with the answers. Goethe spoke for our generation as well as his own when he said, "It is not possible to wait with new explorations until man is a moral being."

Behavioral genetics forces us to confront the fact that we are, in essence, chemical beings—vastly complex, adaptable, unique, and unpredictable to be sure, but chemical nonetheless. We are learning to identify those chemicals, to decipher how they are made, to discover what they do, and to fix them when they are defective. In the process, science, once again, has forever changed the way we think about ourselves and our place in the universe.

As society continues to explore the unanswered questions, we must remain constantly aware of a central challenge facing all scientists. Ultimate success will be measured not so much in the vast range of questions that we will surely learn to answer, but in how society applies its limited resources to discover a path that enlightens, informs, and enriches all people.

The Search for Extraterrestrial Intelligence: Are We Alone in the Universe?

Were we to locate a single extraterrestrial signal, we would know immediately one great truth: that it is possible for a civilization to maintain an advanced technological state and *not* destroy itself.

–Philip Morrison *The Search for Extraterrestrial Intelligence*
(1977)

In the September 19, 1959, issue of the prestigious journal *Nature*, physicists Giuseppe Cocconi and Philip Morrison of Cornell University posed a shocking new question for science. Is there a way, they asked, of conducting a systematic search for space aliens?

The existence of extraterrestrial life, a mainstay of science fiction for more than a century, has been debated on philosophical grounds for thousands of years. Scientific analysis, however, requires more than fanciful speculation; it relies on reproducible observations and experiments. Prior to the late 1950s, before the launch of *Sputnik* and the dawning of the space age, any quest for alien life was mocked as beyond the domain of serious science. Cocconi and Morrison transformed the scientific status of the search for extraterrestrial intelligence, or SETI, by proposing a simple and compelling experiment in a direct and persuasive style. They based their argument on the intriguing possibility that a scientifically advanced alien society, thinking the sun a likely home for intelligent life, might have established a radio signal beacon in the hope of attracting our attention. In powerful prose they concluded: "We therefore feel that a discriminating search for signals deserves a considerable effort. The probability of success is difficult to estimate; but if we never search, the chance of success is zero."

Within two years the National Academy of Sciences had organized a meeting of some of the world's leading astronomers to plot a strategy for SETI research.

WHY IS THE SEARCH
SO COMPELLING?

Nature displays a remarkable economy of design. Every phenomenon that occurs once seems to repeat over and over again. Each individual volcanic eruption, lightning bolt, or snowstorm is unique in detail, but all such events are examples of nature's endlessly recurring patterns.

The solar system features comets, asteroids, planets, and moons in abundance. As we peer outward into deep space, we see countless billions of galaxies, each with countless billions of sunlike stars. And we have every reason to believe that the majority of those suns will have several planets, even though direct observation of all but a few of those worlds is presently beyond the resolution of our most powerful telescopes. By even the most conservative estimates, the universe's inventory of Earthlike planets is beyond imagining.

Those distant planets form from the same kinds of atoms, are governed by the same physical laws, and some of them are likely to display many of the same natural processes as on Earth. Each Earthlike planet will have volcanoes, lightning, and snow. In such a redundant universe as ours, it seems to many scholars inconceivable that life has arisen only once. And if life is common in the universe, then surely we are not the only species who can pose such questions as "Are we alone?"

For much of recorded history, humans celebrated Earth as the center of the universe, and themselves as the Creator's ultimate work. Time after time during the past several centuries, scientific discoveries have challenged this comforting view of human superiority. Nicolas Copernicus argued that the Earth circles the sun; we are not the physical center of the universe. Charles Darwin claimed that humans evolved from earlier animal forms; we are not a special creation. Edwin Hubble demonstrated that our galaxy of 100 billion suns is but one of countless billions of galaxies; our home represents the tiniest fraction of the universe. With each of these epic discoveries, the human role in the cosmos seemed diminished.

Yet, for now, we humans and our home still appear to be unique. Earth is the only known planet with life, and we are its only intelligent, technological species. But we cannot know with certainty that humans alone have developed the technology to look beyond their own world. Until such time as we detect transmissions from a distant, alien be- ing—or perhaps meet them face-to-face—we will never be sure.

First contact with an extraterrestrial intelligence has become a cliché of science fiction novels and movies. Such close encounters are por- trayed alternatively as our species' ultimate nightmare or as an unparal- leled opportunity. Yet one fact remains. The discovery of an alien intel- ligence would, in an instant, transform our perception of humanity's place in the cosmos like no other event we can imagine.

"Are we alone?" is a question unlike most others in science. The great majority of profound scientific questions focus on the origins of our physical world, and the dynamic processes by which it changes. Such scientific understanding evolves gradually, in countless tiny incre- ments. Many deep questions, such as the processes by which life origi- nated or the ultimate fate of the universe, may never be completely answered by humans. The search for other intelligent, technological life-forms differs because it has the simplest possible answer. Are we alone? "Yes, we are" or "No, we are not" provide the only possible answers. The question also differs from most others in science in that it may be answered even if we never lift a finger to study it. They may find us first.

Until we discover an alien intelligence, we can never be certain of our solitude. A single, unambiguous communication or direct contact with another technological species, however, will instantly answer the question, and change forever our perception of life in the universe. Of course, once that alien life is found, myriad new questions about their distant origins, novel biology, technological capabilities, and ultimate intentions will consume scholars' lives for centuries to come.

WHAT ARE THE CHANCES OF FINDING ET?

Though we know of no alien technological societies, we can attempt to guess the probability of their existence. A procedure for making such an estimate was proposed in the 1960s by SETI pioneer Frank Drake, professor of astronomy at the San Diego campus of the University of California. Drake suggested that seven factors, each worthy of independent thought and study, contribute to the probability that intelligent life exists on other worlds.

Factor 1.
How Many Stars Are Sunlike?

Life as we know it cannot evolve and thrive without the energy of a sunlike star. Larger stars, many times the mass of the sun, burn fiercely for less than a billion years and then die. Intelligent life is unlikely to evolve in so short a span. Long-lived stars much smaller than the sun, on the other hand, may not emit sufficient energy to sustain life. Only relatively stable stars about the size of the sun burn long enough, and steadily enough, to nurture the origin and gradual evolution of intelligent life.

Life as we know it also needs a variety of chemical elements. According to current models of the early universe, the first generation of stars was composed almost entirely of hydrogen and helium, which cannot, by themselves, support life. Heavier elements, including the carbon, nitrogen, and oxygen essential for terrestrial life, were produced in the nuclear reactions of first-generation stars, and were distributed throughout the galaxy when the largest of those stars exploded in supernovas. Our sun is a later-generation star, and so the solar system has a ready supply of carbon, oxygen, and life's other essential elements.

Astronomers find the task of searching for sunlike stars relatively easy. They examine the light from stars, one by one, to detect the distinctive yellow-rich glow characteristic of the sun's approximately 6000° C surface temperature. They find that only a small fraction of stars, perhaps only one in a hundred, provides a reasonable match to our sun; most stars are much smaller than our own. Furthermore, more than half of all visible stars are part of binary or multiple star systems, around which stable planetary orbits are (at least in theory) less likely. Nevertheless, our galaxy of 100 billion stars holds perhaps half a billion possible suns.

Can Life Arise Without a Star?

American astronomer Harlow Shapley proposed in the 1960s that life might evolve on objects called "brown dwarfs," planetlike objects that are much larger than the Earth, but not quite large enough to form a star. Such a world would stay warm for billions of years, as its internal heat gradually leaked to the surface. Given sufficient heat energy and the necessary nutrients, there is no reason why life could not begin on such a dark world.

Brown dwarfs do not shine brilliantly in the night sky, and are thus difficult to detect. No one knows for sure how many there are, though many experts suspect their number far exceeds the population of visible stars. A new generation of heat-sensing infrared telescopes is now allowing astronomers to locate and describe these elusive objects for the first time.

Factor 2. How Many Sunlike Stars Have Solar Systems?

In our present view of star formation, planetary systems are almost inevitable. Stars arise from the gravitational collapse of immense clouds of hydrogen and other matter called nebulae. As a nebula contracts, it begins to spin like an ice skater pulling in her arms. Faster and faster the cloud rotates, adopting a flattened disk shape, much like a spinning mass of pizza dough thrown into the air. Most of the nebular material concentrates at the center to form the star, but a thin disk of matter remains in orbit.

Gradually, gravity causes the matter in this disk to clump together, first into pieces no larger than a grain of sand, but then into larger and larger chunks. Bigger pieces sweep up the smaller, until eventually, perhaps in a few thousand years, several dozen objects, each hundreds of miles in diameter, compete for space in a newly formed planetary system. The final stage of this process features epic collisions between moon-sized objects, until a few planets, each well spaced from the others, remain. In thousands of computer-generated simulations of this chaotic process, astronomer George Wetherill of the Carnegie Institution of Washington finds that planets invariably form.

Even the nearest neighboring stars are many trillions of miles away—much too far to see Earthlike planets. As of this writing, only tentative indications of planetary systems outside of our own have been found. Slight wobbly variations in the position, intensity, or velocity of some stars suggests the presence of smaller companion objects, though these observations are made at the extreme limits of telescope sensitivities. But many astronomers would agree: If there's a star, chances are good there are planets. Thus, the Milky Way galaxy could hold half a billion solar systems similar to our own.

Factor 3.
How Many Solar Systems Have Earthlike Planets?

During the nebular process that forms planets, the central star eventually ignites and powerful solar winds blow outward. Lighter gaseous elements, including hydrogen and helium, are swept away from the inner planets. In our own solar system, the inner four planets are thus relatively small, dense, rocky worlds stripped of volatile elements, while the next four are gas giants, rich in hydrogen and helium.

The hallmark of Earth, one of the inner rocky worlds, is the abundance of liquid water, which is stable in the relatively narrow temperature range between 0 and 100 degrees Celsius. Venus, the next nearest planet to the sun, is much too hot, while the more distant Mars is too cold, to allow life-giving rivers and oceans. It seems unlikely, therefore, that any planetary system could boast more than one or two worlds with liquid water.

Nevertheless, in spite of these restrictions, the number of Earthlike planets orbiting sunlike stars in our galaxy may be in the hundreds of millions.

The Tide of Life: Does Life Require a Large Moon?

Liquid water may be a necessary but not sufficient condition to promote life. Some biologists speculate that life began in a coastal tidal zone, where the sun's energy bathed a shallow pool. With each low tide, the sun evaporated water and the pool became more concentrated in the chemical nutrients of life. If this scenario is correct, then moon-induced tides may play a key role in the origin of life.

The Earth's moon is a remarkable object. Most other moons are

relatively small compared to their planet. They either were captu:
the planet's gravity, or they were formed from leftover debris in]
tary orbit. But the Earth's moon is much too large—larger tha
planets Mercury and Pluto, and almost as large as Mars. Scientists speculate that the moon formed about 4.5 billion years ago, at a time when
the Earth and a second, smaller planet competed for our orbital region
of the solar system. After millions of years of jockeying for the same air
space, the two objects collided in an epic event, dubbed the "big
splash." The smaller planet was pulverized along with a giant chunk of
the Earth's mantle, and a great mass of molten rock thrown into orbit
became the moon.

Small gravitationally captured moons are common around solar system planets, and, presumably, around planets in other solar systems as
well. Nevertheless, only a small fraction of all Earthlike planets is likely
to have experienced such a cataclysmic moon-forming event or to experience tides. Thus, if the moon was essential for life's origins, then life
in other planetary systems may be correspondingly rare.

Cosmic Vacuum Cleaners:
Does Life Require a Jupiterlike Planet?

Scientists generally agree that Jupiter, Saturn, and the other giant gas
planets are unlikely places for life to begin—at least life based on carbon and water as we know it. Temperatures are too cold so far from the
sun. But the presence of a giant, Jupiterlike planet may be essential for
life on an Earthlike planet.

Our solar system contains countless millions of asteroids and comets, which are large enough to wreak havoc on a life-sustaining world.
Every few tens of millions of years, one of these objects hits the Earth
and dramatically disrupts the biosphere. The impact that may have
killed the dinosaurs 65 million years ago was neither the most severe
nor the most recent of these catastrophes. Intelligent life on Earth has
arisen during the long interval between such random destructive
events.

The giant planets in our solar system play a vital role in protecting
Earth by acting as gravitational vacuum cleaners that sweep up the vast
majority of these potentially lethal projectiles. Without the gravitational groundskeeping of Jupiter and Saturn, life on Earth would have
experienced many times more destructive collisions, perhaps one every

few ten thousand years. Under such a harsh bombardment, evolution of intelligent life forms wouldn't have stood much of a chance.

Factor 4.
What Fraction of Earthlike Planets Have Life?

Life arose on the only Earthlike planet we know, but scientists are strongly divided on whether life is inevitable on other similar worlds. An articulate and persuasive group of molecular biologists led by Christian de Duve argues that life is a likely outcome of simple chemical reactions that must occur on virtually any planet where liquid water and carbon are abundant. Others, who cite the astonishing complexity of even the simplest single cell, disagree. They suspect that life is extremely rare, if not unique to our planet. Hence, guesstimates of the number of planets with life in our galaxy range from hundreds of millions to just one.

For the foreseeable future, the only places other than Earth where we can search directly for life are objects in the solar system: the moon, nearby planets, asteroids, and comets. The search for extraterrestrial life continues to have a significant impact on the space program of the United States. All deep-space probes are thoroughly sterilized in order to prevent the spread of Earth-born bacteria to other worlds. By the same token, lunar astronauts and their precious rock samples were tightly quarantined for a month upon their return to the Earth in the fear that alien microorganisms might infect our ecosystem. Nothing biological was found in those samples; we now believe the moon is lifeless. The harsh extremes of searing lunar days and frigid lunar nights appear to have sterilized the rugged surface.

Of all the other bodies in the solar system, Mars has long been viewed as the most likely haven for life. The red planet's ice caps, presumed by early observers to imply a ready supply of water, also fueled speculation about the origin of enigmatic linear features on the Martian surface. Italian astronomer Giovanni Schiaparelli described *canali* (literally, "channels") criss-crossing Mars, without implying their artificial origin. Bostonian Percival Lowell, however, equated these features with artificial "canals" and claimed that they formed a sophisticated water-distribution system from the poles to drier equatorial regions. His precise and detailed maps of this Martian infrastructure, completed nearly a century ago, were confirmed by many other distinguished astronomers. The nature and implications of Martian life were discussed in leading scientific publications and conferences of the day. The idea of a superrace of Martians has persisted in popular fiction long after Lowell's wishful evidence for a technological society was discounted.

Even after Earth-based observations disproved the existence of canals and other advanced technologies, many scientists (most notably Carl Sagan) thought simple one-celled life-forms might thrive in the red Martian soil, and these speculations have been strengthened by recent studies of Antarctic meteorites of Martian origin. These rocks contain tiny spherical structures, complex carbon molecules, and other features that provide tantalizing evidence for possible microbial life on Mars. Nevertheless, proof of life on Mars must await visits to the planet itself. Detailed photographs of the Martian surface show immense canyons, river beds, and other features that prove water once flowed in abundance. Most of Mars's water has evaporated into space, but significant amounts remain locked in permafrost beneath the surface. Could primitive life be found near these ancient underground reservoirs?

One of the most closely watched experiments aboard the *Viking 1* Mars lander was a soil test for biological activity. A scoop of Martian dirt was exposed first to water, then to nutrients, in a search for telltale chemical responses. The soil proved highly reactive, but not in a way characteristic of life. Scientists found no signs of life in the bone-dry soil, and most experts agree that the harsh surface of Mars is now sterile. Others, however, remain optimistic that life may persist close to the poles or deep beneath the surface, or that fossil evidence of Martian organisms might be recovered in ancient layers of sediment.

WARNING: About here in this chapter the cautious reader might wish to wear hip boots and carry a shovel. In spite of volumes of fanciful speculation, no one has the answers, and we have precious little useful data on which to base any conclusions.

Factor 5.
How Often Does Life Lead to Intelligence?

Animals that can learn and reason are unusual but by no means unique on Earth. The intelligence of mammals, including whales, dolphins, great apes, and us, is well documented. More surprising is the completely independent development of reasoning ability among octopi, who can learn complex tasks simply by watching other octopi in adjacent tanks. Given that intelligence has evolved more than once on our world, it is not unreasonable to assume similar developments might be typical of life elsewhere.

The evolution of intelligence requires a highly evolved brain and, some researchers argue, an advanced visual system as well. If the ability to reason and learn confers an advantage on an organism, then, given sufficient time, intelligent species should appear on many life-bearing planets. How many is anybody's guess; our galaxy could hold a hundred million planets with intelligent life, or only one.

Factor 6.
Will Intelligent Life-Forms Attempt to Communicate?

Scientists believe that every observer anywhere in the universe will experience the same physical laws. These laws await any intelligent species that seeks them out. Nevertheless, intelligence guarantees neither the desire nor the ability to communicate across the vastness of space. If our planet had been entirely covered by oceans, for example, or if life had never ventured forth on land, there is no guarantee that any species could have developed radio. On Earth, technology advanced along a tortuous but inexorable path that was largely dependent on the taming of fire. Fire provided energy for the smelting of metals, facilitated bench-top chemistry, and enabled much other laboratory experimentation that led to the articulation of natural laws. Could a water-bound civilization have done the same?

Factor 7.
How Long Do Advanced Societies Remain Communicative?

SETI researchers have at least some firsthand data regarding the first six factors that might control the evolution of intelligent life. We know of one sunlike star with an Earthlike planet on which a self-aware, technological civilization arose. What we do not yet know is how long such a civilization can survive before it destroys itself or is destroyed. Will we venture out into space and people this sector of the Milky Way galaxy, à la *Star Trek*? Will we run out of essential resources? Will we lose control of powerful nuclear technologies or technologies we have yet to invent in ways that ensure our destruction? Will some emerging infectious organism kill us all? Or will we be done in by the impact of solar system debris onto the Earth?

A civilization need not end to escape detection by distant eavesdroppers. Radio communication may become obsolete, replaced by vast networks of efficient optical fibers or by some as yet undreamed of technology. Perhaps another world's use of radio will be as fleeting as the telegraph or steam engine in our own history. Or perhaps more cautious alien cultures, wary of extraterrestrial invaders, maintain strict radio silence.

Thomas McDonough, SETI coordinator for the Planetary Society, suggests yet another reason that an advanced civilization might cease its attempts at interstellar communication. "Suppose a way was found to stimulate pleasure centers of the brain through narcotics or electricity, and that life-support equipment allowed beings to remain hooked up in perpetual ecstasy, while robots took care of maintenance," he remarks. "There could be whole planets full of happy, uncommunicative pleasure-addicts." Clearly, the absence of extraterrestrial signals does not prove an absence of intelligent life.

Though they are little more than pure guesswork, estimates of the half-life of a society with radio technology range from a pessimistic few decades to millions of years. If the average alien intelligence transmits radio waves for less than a century before it self-destructs or becomes otherwise engaged, then there is little chance that we will ever hear their brief and futile messages.

What Is the Solution to the Drake Equation?

Large uncertainties are associated with each of the seven factors that comprise the Drake equation. Even if we can be reasonably confident that our galaxy holds a half billion Earthlike planets (factors 1, 2, and 3, combined), we have almost no data on the probability of life (factor 4), intelligence (factor 5), or the behavior of an alien technological civilization (factors 6 and 7). Consequently, combining all seven factors to estimate the number of technological alien societies is a house of cards. Nevertheless, Drake and subsequent scientists have attempted to place upper and lower bounds on the probability of contact.

By his most optimistic estimate, assuming that almost every Earthlike planet eventually develops intelligent life, Drake suggested that signals will originate from one in every four potential planets—perhaps 150 million in all. His most pessimistic calculation assumes signals are broadcast from one in a million Earthlike planets—perhaps only 500 in the entire galaxy.

Astronomer Carl Sagan favored an intermediate number of as many as ten million technological communities. If so, then the chances are good that one of them lives within 100 light-years of our solar system. Our radio waves, first broadcast commercially early this century, may have already begun to bathe that distant planet. And our closest stellar neighbors may have already formulated a reply.

Why Aren't They Here?

Many critics who disagree with Frank Drake and Carl Sagan's most optimistic estimates of millions of other intelligent communities ask one central question: If so many exist, why aren't they here?

After only a few hundred years of technological development, humans are poised to colonize space. Using existing technologies (and a huge investment of resources), we could build giant spaceships and begin the voyage to a nearby star. Gradually, we would spread outward from the solar system, using distant asteroids and comets for raw materials, and finding new planetary homes along the way.

Even making the most conservative assumptions, that no new propulsion technologies are discovered and outward exploration proceeds at a leisurely rate of only a hundredth the speed of light, the entire galaxy 100,000 light-years across could be populated within a few tens of millions of years—a mere instant in the 10-billion-year lifetime of our solar system. If humans could undertake such a venture now, why couldn't any other technological community as well?

How many billions of years ago did the first alien Newton discover the universal laws of motion? How long has it been since the periodic table of elements was proposed on a distant world? Why aren't these ancient alien explorers here? Why haven't we heard from them? Some scientists conclude either that technological societies are extremely rare (perhaps unique), or that it is exceedingly difficult to sustain a viable community for hundreds of years in space. Others suggest that intelligent alien species may be disinterested in or wary of contacting others, or that *all* such societies, without exception, must destroy themselves.

SETI

Unlike many SETI experts, Philip Morrison refuses to speculate on the best estimate for the Drake equation. "That isn't the right question," he claims. "The question is really, 'Should we do something to find out?' . . . To find out you must do something empirical." And so the search goes on.

Countless alien civilizations may be out there, sending their messages of greetings, waiting for our reply. What should we do about it?

Searching the Cosmic Haystack

Radio waves and microwaves, traveling at light speed through space, provide the most practical medium for interstellar communication, while radio telescopes provide ideal receivers for these potential signals. Scientists suspect that a communication from an intelligent species will be unambiguous. Perhaps it will begin with a sequence representing the prime numbers: 1, 2, 3, 5, 7, 11, and 13 short pulses repeated over and over again to get our attention. Then a binary-coded message of dots and dashes could convey basic information about the senders and their world.

The central problem confronting SETI scientists is where and how to search for such a signal. Three radio broadcast variables must be matched: direction to the source, frequency of the signal (as modified by red or blue shifts), and sensitivity (the intensity of the source above background noise). The staggering number of possible combinations of these conditions makes the SETI effort something akin to searching for a needle in a cosmic haystack.

In their original proposal, Cocconi and Morrison suggested that the initial search for ET should focus on seven nearby sunlike stars. They also argued that receivers should be tuned to a distinctive, sharp radio frequency of hydrogen atom emissions that would be known to all advanced civilizations—about 1420 megahertz, corresponding to a wavelength of about 21 centimeters. Stars do not emit much of their energy as radio or microwaves, so even a relatively weak alien signal could stand out like a beacon in the sky. Many SETI efforts, including searches now under way, concentrate on this 1420 megahertz region.

Subsequent SETI efforts have employed a variety of strategies. Pinpoint searches over a wide range of radio frequencies, wide-angle searches over a narrow range of frequencies, broad-area searches for the most powerful transmissions, and narrow-area searches for extremely faint signals have all been attempted. Nor has the exploration been limited to radio transmissions. Advanced civilizations might communicate with lasers or masers tuned to higher frequencies corresponding to short microwaves, infrared, or visible light. Even so, radio waves remain the most likely transmission medium in most experts' minds, because they are much easier to transmit and receive over interstellar distances, and they are inherently less noisy than shorter wavelength radiation.

For more than three decades, radio astronomers have watched the skies. From time to time, distinctive signals have been found. In 1960, Frank Drake initiated the first systematic search for radio signals from nearby Earthlike stars—an effort whimsically dubbed Project Ozma. Working with the 85-foot-diameter dish of the National Radio Astronomy Observatory in Green Bank, West Virginia, Drake detected a strong eight-pulse-per-second signal coming from Epsilon Eridani, one of Cocconi and Morrison's seven target stars. After an initial reaction of shock and euphoria, Drake and coworkers set about the exacting task of documenting their discovery. Unfortunately, the source moved, and then it disappeared. After weeks of checking, they realized that it was nothing more than man-made interference, perhaps an experimental radar jamming signal.

Irish astronomer Jocelyn Bell enjoyed a similar experience in 1967, when she detected a monotonous pulsating noise with a recently constructed radio telescope at Cambridge University in England. Unlike Drake's experience, the source was steady and pinpointed, and the British team, led by Anthony Hewish, soon identified three more pulsating sources. These "pulsars" were eventually shown to be rapidly rotating remnants of exploded stars; undaunted astronomers still gave them the playful catalogue designation LGM, for little green men.

The search goes on. While the government-sponsored portion of the SETI radio search ended in 1993, private funding to the tune of more than ten million dollars annually supports the effort. The most extensive study, a direct continuation of NASA-supported research of the early 1990s, concentrates on specific likely target stars, monitoring each with thousands of channels. Computers engage in the epic task of pattern-recognition.

Smaller-scale projects are now attempting to survey the entire sky using the narrow hydrogen frequency range. The Planetary Society supports one such effort, Project Meta, masterminded by physicist Paul Horowitz of Harvard University. Horowitz conceived and constructed the largely automated, full-time radio survey that simultaneously monitors 8.4 million finely tuned channels. Project Phoenix, a complementary effort begun by astrophysicist Jill Tarter, recently spent several months scanning the southern skies from radio telescopes in Australia.

For more than two decades, radio astronomers John Kraus and Robert Dixon of Ohio State University have engaged in a similar investigation, project Cosmic Search, using a giant radio telescope in a Delaware, Ohio, field. When Kraus's government support for conventional radio astronomy dried up in 1972, Dixon suggested that their Ohio facility was ideally suited to investigate signals near the 1420 megahertz frequency. In subsequent years, Kraus has channeled much of his own considerable income from antenna designs and reference books to maintain the telescope. The Ohio facility has devoted many thousands of hours to the SETI effort. Apart from one intense, enigmatic 37-second signal in August 1977, no hint of an extraterrestrial communication has been found.

Have Humans Already Made Contact?

Each year hundreds of Americans claim to have seen alien spacecraft, or even to have met intelligent outer-space creatures face-to-face. Bookstores feature dozens of titles on flying saucers, while television offers show after show on bizarre alien abductions. Yet, in spite of numerous claims of firsthand experiences, most scientists remain wary of reports of unidentified flying objects. The entire subject has been relegated to the category of pseudoscience, along with astrology, extrasensory perception, and crystal power.

Why do scientists treat these reported observations with such skepticism? Are they close-minded? What would constitute proof of alien contact?

UFOs are regarded with skepticism by science for two important reasons. First, science deals with recurrent phenomena, tangible evidence, and reproducible measurements. UFOs are, by their very ephemeral and unpredictable nature, impossible to study in a systematic way. Sightings of UFOs deserve rigorous and thoughtful analysis, but anecdotal evidence alone cannot serve as scientific data.

Second, scientists always search for the most likely explanation for any specific observation or phenomenon—the principle of Occam's razor. While any UFO sighting *might* be the result of an extraterrestrial visit, there always seem to be more likely explanations based on what we know about atmospheric phenomena and human nature.

That being said, what are the chances that a technologically advanced civilization will try to contact us? If Carl Sagan's optimistic estimate of 10 million advanced civilizations in our galaxy is correct, then the chances are reasonable that one of them lives within fifty light-years of Earth. Our period of radio brilliance began eighty years ago, so distant aliens have now had ample opportunity to detect and analyze our inadvertent chatter. (It's a sobering thought that Howard Stern and Rush Limbaugh may become homing beacons to aliens of advanced intelligence.)

If, on the other hand, technological civilizations are rare at any point in the galaxy's history, then our chances of hearing from one is slim. Furthermore, any distant signals that we detect will have been traveling for centuries. The civilization may well have disappeared long before we learn of it.

What Will ET Look Like?

Of one thing we can be sure. An intelligent extraterrestrial species will almost certainly not look human. Our size, shape, and structure evolved from a cascade of innumerable chance events over hundreds of millions of years—a sequence that is not likely to be reproduced exactly, not even on an identical planet.

We can point to life on Earth, however, to identify features that have evolved many different times. All advanced land and sea animals have eyes, and they obtain a large part of their sensory input from vision. Fossil evidence and existing organisms demonstrate that eyes have evolved independently dozens of times on Earth, in close to a dozen distinctive forms: multifaceted compound eyes, pinhole eyes, eyes with lenses and eyes without. Nevertheless, virtually all organisms have at least two eyes, a requirement for effective depth perception, so eyes in pairs seems a logical evolutionary pattern.

Eyes are most effective when located near the front of an animal, and they are most protected when situated a short distance from the brain that interprets the visual stimuli. It seems likely, therefore, that most aliens will have a recognizable headlike structure. Herbivores such as horses and rabbits often have eyes on opposite sides of their head to provide close to 360 degrees of vision. Predators such as eagles and humans, on the other hand, have eyes facing the same direction for optimum acuity and stereoscopic vision.

Development of technology requires manipulating objects; hence some array of arms or tentacles with handlike grasping ability seems a prerequisite for any alien intelligence. A compact body plan that protects vital organs and a means of locomotion add further constraints to the appearance of any alien creature. Given these requirements, exobi-ologists predict that alien visitors might have body plans in some way analogous to our own. While such creatures almost certainly will not appear humanoid, some of the more exotic latex creations of *Star Wars* and *Star Trek* may not be so far from what a real extraterrestrial might look like.

Whether or not space aliens exist, whether or not they resemble us, the search for extraterrestrial intelligence is likely to remain at the fringes of scientific research. Why should such a fundamental and potentially answerable question receive so little support?

For many scientists the answer may lie in the lack of recognition and prestige afforded negative results. SETI pioneers have spent decades finding absolutely nothing. That result itself undoubtedly has some significance, but what? Perhaps we just haven't looked hard enough or in the right way. Scientists and funding agencies are no doubt wary of spending too much effort on a project that may yield decades more of nothing.

For others, scientists and nonscientists alike, a halfhearted search for ET may also reflect deep-seated misgivings about what might be out there. Even if an alien community is benign, are we ready to learn that we represent but one of countless intelligent forms, and a not very advanced one at that?

ADDITIONAL
READING

Prologue—The Nature of Questions

Feyerabend, Paul. *Against Method: Outline of an Anarchistic Theory of Knowledge*. London: Verso, 1975.

Horgan, John. *The End of Science*. Reading, Mass.: Addison-Wesley, 1996.

Keller, Evelyn Fox. *A Feeling for the Organism: The Life and Work of Barbara McClintock*. New York: W. H. Freeman, 1983.

McPhee, John. *The Control of Nature*. New York: Farrar Straus and Giroux, 1989.

Sonnert, Gerhard. *Gender Differences in Science Careers: The Project Access Study*. Rutgers, New Jersey: Rutgers University Press, 1995.

Chapter 1—Dark Matter

Bartusiak, Marcia. *Through a Universe Darkly: A Cosmic Tale of Ancient Ethers, Dark Matter, and the Fate of the Universe*. New York: HarperCollins, 1993.

Morris, Richard. *Cosmic Questions: Galactic Halos, Cold Dark Matter, and the End of Time*. New York: John Wiley and Sons, 1993.

Overbye, Dennis. "Weighing the Universe." *Science* 272, June 7, 1996, p. 1426.

Riordan, Michael, and Donald Schramm. *The Shadows of Creation: Dark Matter and the Structure of the Universe*. New York: W. H. Freeman, 1994.

Trefil, James. *The Dark Side of the Universe*. New York: Anchor Books, 1989.

Tremaine, Scott. "The Dynamical Evidence for Dark Matter." *Physics Today*, February 1992, pp. 28–36.

Trimble, Virginia. "Existence and Nature of Dark Matter in the Universe." *Annual Review of Astronomy and Astrophysics* 25, 1987, pp. 425–472.

Chapter 2—Fate

Davies, Paul. *The Last Three Minutes: Conjectures about the Ultimate Fate of the Universe*. New York: Basic Books, 1994.

Dressler, Alan. *Voyage to the Great Attractor: Exploring Intergalactic Space*. New York: Knopf, 1994.

Goldsmith, Donald. *Einstein's Greatest Blunder? The Cosmological Constant and Other Fudge Factors in the Physics of the Universe*. Cambridge: Harvard University Press, 1995.

Gribbon, John. *The Omega Point: The Search for the Missing Mass and the Ultimate Fate of the Universe*. New York: Bantam Books, 1988.

Harrison, Edward R. *Cosmology: The Science of the Universe*. Cambridge: Cambridge University Press, 1981.

Hubble, Edwin. *The Realm of the Nebulae*. New Haven: Yale University Press, 1982.

Morris, Richard. *Cosmic Questions: Galactic Halos, Cold Dark Matter, and the End of Time*. New York: John Wiley and Sons, 1993.

Rowan-Robinson, Michael. *The Cosmic Distance Ladder: Distance and Time in the Universe*. New York: W. H. Freeman, 1985.

Trefil, James. *The Moment of Creation*. New York: Macmillan, 1984.

Weinberg, Steven. *The First Three Minutes: A Modern View of the Origin of the Universe*. New York: Basic Books, 1988.

Silk, Joseph. *The Big Bang*. New York: W. H. Freeman, 1989.

Chapter 3—Perfect Symmetry

Hawking, Stephen W. *A Brief History of Time: From the Big Bang to Black Holes*. New York: Bantam Books, 1988.

Mukerjee, Madhusree. "Explaining Everything," *Scientific American*, January 1996, pp. 88–94.

Ne'eman, Yuval, and Yoram Kirsh. *The Particle Hunters*. Cambridge: Cambridge University Press, 1986.

Pagels, Heinz R. *Perfect Symmetry: The Search for the Beginning of Time.* New York: Simon and Schuster, 1985.

Peat, F. David, *Superstrings and the Search for the Theory of Everything.* Chicago: Contemporary Press, 1988.

Riordan, Michael. *The Hunting of the Quark: A True Story of Modern Physics.* New York: Touchstone, 1987.

Chapter 4—Stuff

Aldersey-Williams, Hugh. *The Most Beautiful Molecule: The Discovery of the Buckyball.* New York: John Wiley and Sons, 1995.

Amato, Ivan. *Stuff: The Materials the World Is Made Of.* New York: Basic Books, 1997.

Atkins, Peter. *The Periodic Kingdom: A Journey into the Land of the Chemical Elements.* New York: Basic Books, 1995.

Cotterill, Rodney. *The Cambridge Guide to the Material World.* Cambridge: Cambridge University Press, 1985.

Gordon, J. E. *The New Science of Strong Materials: Or Why You Don't Fall Through the Floor.* Princeton: Princeton University Press, 1976.

Hazen, Robert. *The Breakthrough: The Race for the Superconductor.* New York: Summit Books, 1988.

Hazen, Robert. *The New Alchemists: Breaking Through the Barriers of High Pressure.* New York: Times Books, 1994.

Hoffmann, Roald. *The Same and Not the Same.* New York: Columbia University Press, 1995.

Holden, Alan. *The Nature of Solids.* New York: Dover Publications, 1992.

Meikle, Jeffrey. *American Plastic: A Cultural History.* New Brunswick, N.J.: Rutgers University Press, 1995.

von Baeyer, Hans Christian. *Taming the Atom: The Emergence of the Visible Microworld.* New York: Random House, 1992.

Chapter 5—The Quest for Energy

American Chemical Society, "Energy." *Chemical and Engineering News,* 69 (special issue), June 17, 1991, pp. 18–46.

Cordey, J. Geoffrey, Robert J. Goldston, and Ronald R. Parker, "Progress Toward a Tokamak Fusion Reactor." *Physics Today*, January 1992, pp. 22–30.

Davis, Jed R., and others. "Energy for Planet Earth." *Scientific American* 263 (special issue), September 1990, pp. 54–163.

Fowler, John M. *Energy and the Environment* (2nd edition). New York: McGraw-Hill, 1984.

Furth, Harold P. "Fusion." *Scientific American* 273, September 1995, pp. 174–176.

Hagan, William J., Roger Bangerter, and Gerald L. Kulcinski. "Energy from Inertial Fusion." *Physics Today*, September 1992, pp. 42–50.

Hazen, Margaret, and Robert Hazen. *Keepers of the Flame*. Princeton: Princeton University Press, 1992.

Herman, Robert. *Fusion: The Search for Endless Energy*. Cambridge: Cambridge University Press, 1990.

Hoagland, William. "Solar energy." *Scientific American* 273, September 1995, pp. 170–173.

Hubbard, Harold M., Paul Notari, Satyen Deb, and Shimon Awerbach. *Progress in Solar Energy Technologies and Applications*. Boulder, Colorado: American Solar Energy Society, 1994.

Johansson, Thomas B., and others, Eds., *Renewable Energy: Sources for Fuels and Electricity*. Washington, D.C.: Island Press, 1993.

Lindl, John D., Robert L. McCrory, and E. Michael Campbell, "Progress Toward Ignition and Burn Propagation in Inertial Confinement Fusion." *Physics Today*, September, 1992, pp. 32–40.

National Research Council, *Plasma Science: From Fundamental Research to Technological Applications*. Washington: National Academy Press, 1995.

Yergin, Daniel. *The Prize: The Epic Quest for Oil, Money, and Power*. New York: Simon and Schuster, 1991.

Zweibel, Kenneth, and Paul Hersch. *Basic Photovoltaic Principles and Methods*. New York: Van Nostrand Reinhold, 1984.

Chapter 6–Core Knowledge

Brown, G. C., and A. E. Mussett. *The Inaccessible Earth*. London: George Allen and Unwin, 1981.

McPhee, John. *Basin and Range*. New York: Farrar Straus and Giroux, 1981.

McPhee, John. *In Suspect Terrain*. New York: Farrar Straus and Giroux, 1983.

McPhee, John. *Rising from the Plains*. New York: Farrar Straus and Giroux, 1984.

National Research Council, *Mount Rainier: Active Cascade Volcano*. Washington: National Academy Press, 1994.

Parker, Ronald. *Inscrutable Earth*. New York: Scribner's, 1984.

Press, Frank, and Raymond Siever. *Understanding Earth*. New York: W. H. Freeman and Co., 1994.

Redfern, Martin. *Journey to the Centre of the Earth: The New Geology*. London: BBC Press, 1991.

Sullivan, Walter. *Continents in Motion*. New York: American Institute of Physics Press, 1991.

Vogel, Shawna. *Naked Earth; The New Geophysics*. New York: Dutton, 1995.

Weiner, Jonathan. *Planet Earth*. New York: Bantam Books, 1986.

Wysession, Michael. "The Inner Workings of the Earth." *American Scientist* 83, March–April 1995, pp. 134–147.

Chapter 7–The Fate of the Earth

Benedick, Richard. *Ozone Diplomacy: New Directions in Safeguarding the Planet*. Cambridge: Harvard University Press, 1991.

Commoner, Barry. *Making Peace with the Planet*. New York: Pantheon, 1990.

Crowley, Thomas J., and Gerald R. North. *Paleoclimatology*. New York: Oxford University Press, 1991.

Davis, Kingsley, and Mikhail S. Bernstam. *Resources, Environment, and*

Population: Present Knowledge, Future Options. New York: The Population Council and Oxford University Press, 1991.

Erlich, Paul. *The Population Bomb* (revised edition). New York: Ballantine Books, 1986.

Gallagher, Richard, and others. "Big Questions for a Small Planet." *Science,* July 21, 1995 (special issue), pp. 283, 313–360.

McPhee, John. *Control of Nature.* New York: Farrar Straus and Giroux, 1989.

Michaels, Patrick. *The Sound and the Fury: The Science and Politics of Global Warming.* Washington, D.C.: Cato Institute, 1992.

Moffett, George D. *Critical Masses: The Global Population Challenge.* New York: Viking, 1994.

Phillips, Kathryn. *Tracking the Vanishing Frogs: An Ecological Mystery.* New York: Penguin, 1994.

Ray, Dixie Lee. *Trashing the Planet.* Washington, D.C.: Regnery Gateway, 1990.

Schneider, Stephen, *Global Warming: Are We Entering the Greenhouse Century?* San Francisco: Sierra Club Books, 1989.

Zimmerman, Michael. *Science, Nonscience, and Nonsense: Approaching Environmental Literacy.* Baltimore: The Johns Hopkins University Press, 1995.

Chapter 8—Origins

de Duve, Christian. *Vital Dust: Life As a Cosmic Imperative.* New York: Basic Books, 1995.

de Duve, Christian. "The Beginnings of Life on Earth." *American Scientist* 83, (September–October 1995), pp. 428–438.

Deamer, D. W., and G. L. Fleischaker, Eds. *Origins of Life: The Central Concepts.* Boston: Jones and Bartlett, 1994.

Folsome, C. E. *The Origin of Life: A Warm Little Pond.* San Francisco: W. H. Freeman, 1979.

Gesteland, Raymond E., and John F. Atkins, Eds. *The RNA World.* Cold Spring Harbor, N.Y.: Cold Spring Harbor Laboratory Press, 1993.

Harold, F. M. *The Vital Force: A Study of Bioenergetics.* New York: W. H. Freeman, 1986.

Kauffman, Stuart. *At Home in the Universe: The Search for the Laws of Self-Organization and Complexity*. New York: Oxford University Press, 1995.

Monod, Jacques. *Chance and Necessity*. New York: Knopf, 1971.

Morowitz, Harold. *Beginnings of Cellular Life: Metabolism Recapitulates Biogenesis*. New Haven: Yale University Press, 1992.

Radetsky, Peter. "How Did Life Start?" *Discover* 13, November 1992, pp. 74–82.

Schopf, J. W., Ed. *Major Events in the History of Life*. Boston: Jones and Bartlett, 1992.

Shapiro, Robert. *Origins: A Skeptic's Guide to the Creation of Life on Earth*. New York: Summit, 1986.

Chapter 9—The Language of Life

Berg, Paul, and Maxine Singer. *Dealing with Genes: The Language of Heredity*. Mill Valley, California: University Science Books, 1992.

Cavalieri, Liebe F. *The Double-Edged Helix: Science in the Real World*. New York: Columbia University Press, 1981.

Crick, Francis. *What Mad Pursuit*. New York: Basic Books, 1988.

Davern, C. I., Ed. *Genetics*. New York: W. H. Freeman, 1985.

Freifelder, D., Ed. *Recombinant DNA*. San Francisco: W. H. Freeman, 1978.

Gonick, Larry, and Mark Wheelis. *The Cartoon Guide to Genetics*. New York: HarperCollins, 1991.

Judson, Horace Freeland. *The Eighth Day of Creation*. New York: Simon and Schuster, 1979.

Kevles, Daniel J., and Leroy Hood, Eds. *The Code of Codes: Scientific and Social Issues in the Human Genome Project*. Cambridge: Harvard University Press, 1992.

Koshland, Daniel E., Ed. *Biotechnology: The Renewable Frontier*. Washington, D.C.: American Association for the Advancement of Science, 1986.

Nelkin, Dorothy, and M. Susan Linde. *The DNA Mystique: The Gene as a Cultural Icon*. New York: W. H. Freeman, 1995.

Sayre, Anne. *Rosalind Franklin & DNA*. New York: W. W. Norton, 1975.

Watson, James. *The Double Helix*. New York: Athenaeum, 1985.

Chapter 10—Evolution

Clube, S.V.M., Ed. *Catastrophes and Evolution: Astronomical Foundations*. Cambridge: Cambridge University Press, 1989.

Dawkins, R. *The Selfish Gene*. New York: Oxford University Press, 1989.

Erlich, Paul, and Edward Wilson. "Biodiversity Studies: Science and Policy." *Science* 253, August 16, 1991, pp. 758–762.

Gould, Stephen Jay. *Wonderful Life*. New York: Norton, 1989.

Gould, Stephen Jay, and Niles Eldredge, "Punctuated Equilibrium Comes of Age." *Nature* 366, November 18, 1993, pp. 223–228.

Johanson, D. C., and M. Edey. *Lucy: The Beginnings of Humankind*. London: Granada, 1981.

Leakey, Richard, and Richard Lewin. *Origins Reconsidered*. New York: Doubleday, 1992.

Nilsson, Dan-E., and Susanne Pelger. "A pessimistic estimate of the time required for an eye to evolve." *Proceedings of the Royal Society of London B* 256, 1994, pp. 53–58.

Raup, David. *The Nemesis Affair: A Story of the Death of Dinosaurs and the Ways of Science*. New York: Norton, 1986.

Raup, David. *Extinction: Bad Genes or Bad Luck*. New York: Norton, 1992.

Rudwick, Martin. *The Meaning of Fossils: Episodes in the History of Paleontology*. New York: Neale Watson Academic Publications, 1976.

Schopf, J. W., Ed. *Major Events in the History of Life*. Boston: Jones and Bartlett, 1992.

Stanley, Steven. *Extinction*. New York: Scientific American Books, 1987.

Stanley, Steven. *Earth and Life Through Time* (second edition). New York: W. H. Freeman, 1989.

Ward, Peter. *The End of Evolution: On Mass Extinction and the Preservation of Biodiversity*. New York: Bantam Books, 1994.

Weiner, Jonathan. *The Beak of the Finch*. New York: Knopf, 1994.

Wilson, Edward O. *The Diversity of Life*. Cambridge: Harvard University Press, 1992.

Chapter 11—Growing Up, Growing Old

Angier, Natalie. *Natural Obsession: The Search for the Oncogene*. Boston: Houghton-Mifflin, 1988.

From Egg to Adult. Bethesda, Md.: Howard Hughes Medical Institute, Report No. 3, 1992.

Hines, Pamela J., and others. "Frontiers in Development." Special issue of *Science* 266, October 28, 1994, pp. 523, 561–614.

Lawrence, Peter. *The Making of a Fly: The Genetics of Animal Design*. Oxford: Blackwell, 1992.

Nilsson, Lennart, and Lars Hamberger. *A Child Is Born*. New York: Delacorte Press, 1990.

Takahashi, Joseph S., and Michelle Hoffman. "Molecular Biological Clocks." *American Scientist* 83, 1995, pp. 158–165.

Varmus, Harold, and Robert A. Weinberg. *Genes and the Biology of Cancer*. New York: Scientific American Library, 1993.

Wolpert, Lewis. *The Triumph of the Embryo*. New York: Oxford University Press, 1991.

Chapter 12—The Human Brain

Chalmers, David J. "The Puzzle of Conscious Experience." *Scientific American*, December 1995, pp. 80–86.

Churchland, Paul M. *The Engine of Reason, the Seat of the Soul*. Cambrige: MIT Press, 1995.

Crick, Francis. *The Astonishing Hypothesis: The Scientific Search for the Soul*. New York: Scribner's, 1994.

Edelman, Gerald. *The Remembered Present: A Biological Theory of Consciousness*. New York: Basic Books, 1994.

Edelman, Gerald. *Bright Air, Brilliant Fire: On the Matter of the Mind*. New York: Basic Books, 1992.

Gelernter, David. *The Muse in the Machine: Computerizing the Poetry of Human Thought*. New York: Free Press, 1994.

Greenfield, Susan. *Journey to the Centers of the Mind: Toward a Science of Consciousness.* New York: Freeman, 1995.

Hellige, Joseph B. *Hemispheric Asymmetry: What's Right and What's Left.* Cambridge: Harvard University Press, 1993.

Johnson, George. *In the Palaces of Memory: How We Build the Worlds Inside Our Heads.* New York: Knopf, 1991.

LeVay, Simon. *The Sexual Brain.* Cambridge: MIT Press, 1993.

Penrose, Roger. *Shadows of the Mind: A Search for the Missing Science of Consciousness.* New York: Oxford University Press, 1994.

Restak, Richard. *The Modular Brain: How New Discoveries in Neuroscience Are Answering Age-Old Questions about Memory, Free-Will, Consciousness, and Personal Identity.* New York: Scribner's, 1994.

Rosenfeld, Israel. *The Strange, the Familiar, and the Forgotten: An Anatomy of Consciousness.* New York: Vintage Books, 1994.

Scott, Alwyn. *Stairway to the Mind: The Controversial New Science of Consciousness.* New York: Springer-Verlag, 1995.

Searle, John R. "The Mystery of Consciousness." *New York Review of Books,* November 2, 1995, pp. 60–66; November 16, 1995, pp. 54–61.

Springer, Sally, and Georg Deutsch, *Left Brain, Right Brain* (4th edition). New York: W. H. Freeman, 1993.

Chapter 13—Behavioral Genetics

Bouchard, T. J., Jr., and P. Propping, Eds. *Twins As a Tool of Behavior Genetics.* New York: John Wiley and Sons, 1993.

Dunn, J., and Robert Plomin. *Separate Lives: Why Children in the Same Family Are So Different.* New York: Basic Books, 1990.

Eaves, Lindon J., H. J. Eysenck, and N. G. Martin. *Genes, Culture, and Personality.* New York: Academic Press, 1989.

Gould, Stephen Jay. *The Mismeasure of Man.* New York: Norton, 1983.

Joyson, R. B. *The Burt Affair.* New York: Academic Press, 1990.

Loehlin, J. C. *Genes and Environment in Personality Development.* Newbury Park, California: Sage, 1992.

Nelkin, Dorothy, and M. Susan Lindee. *The DNA Mystique: The Gene As a Cultural Icon.* New York: W. H. Freeman, 1995.

Plomin, Robert, John C. DeFries, and Gerald E. McClearn. *Behavioral Genetics: A Primer* (2nd edition). New York: W. H. Freeman, 1990.

Plomin, Robert, and Gerald E. McClearn, Eds. *Nature, Nurture, and Psychology*. Washington, D.C.: American Psychological Association, 1993.

Wiesel, Torsten N., and others. "Genetics and Behavior." *Science* 264 (special issue), June 17, 1994, pp. 1647, 1685–1739.

Chapter 14—The Search for Extraterrestrial Intelligence

Davies, Paul. *Are We Alone? Philosophical Implications of the Discovery of Extraterrestrial Life*. New York: Basic Books, 1995.

Drake, Frank, and Dava Sobel. *Is Anyone Out There? The Scientific Search for Extraterrestrial Life*. New York: Delacorte Press, 1992.

Friedlander, Michael. *At the Fringes of Science*. Boulder, Colorado: Westview Press, 1995.

Heidmann, Jean, *Extraterrestrial Intelligence*. Cambridge, England: Cambridge University Press, 1995.

McDonough, Thomas R. *The Search for Extraterrestrial Intelligence: Listening for Life in the Cosmos*. New York: John Wiley and Sons, 1987.

Morrison, Philip, John Billingham, and John Wolfe, Eds. *The Search for Extraterrestrial Intelligence*. New York: Dover Books, republication of 1977 NASA document.

Shklovskii, I. S., and Carl Sagan. *Intelligent Life in the Universe*. New York: Delta Books, 1966.

Sullivan, Walter. *We Are Not Alone* (revised edition). New York: Plume, 1994.

Swift, David. *SETI Pioneers: Scientists Talk About Their Search for Extraterrestrial Intelligence*. Tucson: University of Arizona Press, 1990.

ACKNOWLEDGMENTS

During the writing of this book we have benefitted from the advice and expertise of numerous individuals engaged in forefront scientific research. Our special thanks go to Richard Klausner, who helped to outline the scope and content of the book, provided detailed comments on many of the chapters, and was to have been a coauthor before his appointment to the demanding directorship of the National Cancer Institute.

We are grateful to Margaret Hazen and Harold Morowitz (George Mason University) for comprehensive reviews of the entire manuscript. We have also received in-depth reviews of specific chapters by many experts, especially our colleagues at the Carnegie Institution of Washington, including Alan Boss, Ronald Cohen, Pamela Conrad, Patricia Craig, Robert Downs, Russell Hemley, John Graham, Iris Inbar, Kathleen Kingma, David Mao, Frank Press, Charles Prewitt, Vera Rubin, Paul Silver, and David Teeter. Other chapters were reviewed by Stephen Jay Gould (Harvard University), Raymond Jeanloz (Berkeley), Allison Macfarland (George Mason University), Evans Mandes (George Mason University), and James Trefil (George Mason University).

We have received valuable advice and assistance from many researchers, including Scott Albright (Golden Photon, Golden, Colorado), Marvin Cohen (Berkeley), Walter Greenberg (University of Maryland), Dan-Eric Nilsson (University of Lund, Sweden), Robert Park (University of Maryland), Gene Ralph (Applied Solar Energy, City of Industry, California), Nicholas Reiter (Solar Cells, Inc., Toledo, Ohio), Steve Rubin (National Renewable Energy Laboratory), and Frederick Suppe (University of Maryland).

ABOUT THE AUTHORS

ROBERT M. HAZEN is a staff scientist at the Carnegie Institution of Washington's Geophysical Laboratory, and Robinson Professor of Earth Science at George Mason University. Among the books he's authored are *Science Matters: Achieving Scientific Literacy* and *The Breakthrough: The Race for the Superconductor*. He lives in Glen Echo, Maryland. **MAXINE SINGER,** a former chief of the Laboratory of Biochemistry at the National Cancer Institute, is currently president of the Carnegie Institution. She lives in Washington, D.C.